Anonymous

Specifications of Work and Materials Required in the Erection and Completion

Of a New County Court House...

Anonymous

Specifications of Work and Materials Required in the Erection and Completion
Of a New County Court House...

ISBN/EAN: 9783337270933

Printed in Europe, USA, Canada, Australia, Japan

Cover: Foto ©berggeist007 / pixelio.de

More available books at **www.hansebooks.com**

SPECIFICATIONS

OF

WORK AND MATERIALS

REQUIRED IN THE

ERECTION AND COMPLETION

OF A NEW

COUNTY COURT HOUSE,

HEATING, LIGHTING AND POWER PLANT

AND

TUNNEL,

IN THE

CITY OF FORT WAYNE,

ALLEN COUNTY,

INDIANA.

1896.
THE JOURNAL CO., PRINTERS,
FORT WAYNE, IND.

List of Drawings.

No. 24.—Ground Floor Tile Plan.

No. 25.—Office Floor Tile Plan.

No. 26.—Court Room Floor Tile Plan.

No. 27.—Designs for Floor Tile.

No. 28.—Ground Story Ceiling Plan.

No. 29.—Office Story Ceiling Plan.

No. 30.—Court Room Story Ceiling Plan.

No. 31.—Plumbers' Risor Diagram.

No. 32.—Basement Plumbing and Lighting Plan.

No. 33.—Ground Floor Plumbing and Lighting Plan.

No. 34.—Office Floor Plumbing and Lighting Plan.

No. 35.—Court Room Floor Plumbing and Lighting Plan.

No. 36.—Attic Plumbing and Lighting Plan.

No. 37.—Heating Risor Diagram.

No. 38.—Basement Heating Plan.

No. 39.—Basement Ventilating Plan.

No. 40.—Ground Floor Heating and Ventilating Plan.

No. 41.—Office Floor Heating and Ventilating Plan.

No. 42.—Court Room Floor, Heating and Ventilating Plan.

No. 43.—Attic Heating and Ventilating Plan.

No. 44.—Roof Plan.

No. 45.—Foundation Plan.

No. 46.—Ground Floor Plan of Iron Work.

No. 47.—Office Floor Plan of Iron Work.

No. 48.—Court Room Plan of Iron Work.

No. 49.—Attic Floor Plan of Iron Work.

No. 50.—Ceiling Plan of Iron Work.

No. 51.—Roof Plan of Iron Work.

No. 52.—Attic Spandrel Sections.

No. 53.—Details of Roof, and Groined Ceilings.

No. 54.—Details of Roof Trusses.

No. 55.—Details of Inner Dome of Tower.

No. 56.—Details of Outer Dome of Tower.

No. 57.—Main Entrance Stone Work.

No. 58.—Plan of Entrances, Columns, Etc.

General Instructions and Conditions

All conditions and requirements are imperative and apply to each and every portion and branch of the work with equal force.

The party herein designated as the "contractor" shall be held to mean either any contractor, firm of contractors, any member of a firm of contractors, or any corporation contracting for any portion of the work herein specified.

In order that the contractors may know the class and grade of material and apparatus which is desired, specific makes and grades of material and apparatus are specified in different cases.

Where any specific make or grade of material or apparatus is specified, it is done to indicate that the particular make or grade so specified is desired in preference to any other make or grade, and is, in the opinion of the Architect, the best for the purpose of which he has knowledge, but it is to be understood that the Commissioners and Architect will accept any other make or grade which, in their opinion, may be equally good and acceptable.

In case any guarantees are made by the contractor and

not realized, he will be given such time as may be deemed ample in the opinion of the Commissioners and Architect to make good such guarantees, and in case of the contractor failing to comply within the time so limited, the Commissioners reserve the right to reject and refuse payment on such apparatus, or appurtenances, or material that do not meet the guarantees as made, and to hold the same subject to the contractor's order.

The contractors shall furnish the Commissioners with guarantees of his sub-contractors, which guarantees shall be taken for the benefit of the Board of Commissioners also, and may be enforced by said board against each sub-contractor also.

All royalties for patents or any improvements thereof that may be involved in the material or apparatus furnished, or in the use thereof shall be sustained by the contractor, who shall satisfy all demands that may be made at any time for such; and the contractor shall be liable for any damage resulting through suits, royalties or claims of any kind whatsoever, and that any loss or damage to the county through such suits or claims, will be made good by the contractor.

The contractor shall observe and comply in every respect with all the state laws and city ordinances governing building operations, and be fully responsible for each and every violation of the same.

The contractor shall give the proper authorities all requisite notices relating to the work, obtain all official permits and licenses that may be necessary, and pay all charges and fees for the same, and all expenses for water, gas and sewer connections and taxes for use of same.

The contractor shall be held liable and fully responsible for all penalties and damage to life, limb and property that may occur, resulting from any act or negligence of himself, agent, sub-contractor or employee, during the progress of the work, or caused by any operations connected with the work, and shall fully protect and save harmless the said Board of Commissioners, and said county and city from all loss and damages resulting from such act or negligence, or accruing in any way from the doing of the work herein provided for.

The contractor shall be held responsible for the obstruction of streets and pavements, either by tearing up of the same or the accumulation of rubbish or materials, and shall carefully repair and make good all damages at his own expense.

The contractor shall also pay for all necessary surveying and leveling.

The contractor shall, in all cases, be required to use proper care and diligence in bracing, securing and protecting all parts of the work against wind, rain, frosts or accidents or injury of any kind, so far as they may interfere with the stability and perfection of the work, for which there must be no extra charge, and in all cases to use his own judgment as to the amount of care and diligence required.

The contractor shall provide and maintain all requisite guards, lights, and danger signals, and take every precaution to prevent injury to persons.

The contractor is to provide temporary doors and windows and keep the building heated and pay for all fuel and other necessary expenses.

He shall also be required to keep faithful and reliable watchmen on the work at all hours.

The contractor shall give the work his personal attention at all times, or shall place it in charge of competent and reliable foreman, who must be acceptable to the Commissioners and Architect, be constantly present on the work until it is fully completed, and authorized by the contractor to act for him and carry out the directions of the Commissioners and Architect or their representatives.

The Architect reserves the right to require the discharge of any employee whom he may deem incompetent, careless or detrimental in any manner whatever.

The contractor shall not assign the contract; nor shall he sub-let any part of the work without the written consent of the Commissioners and Architect.

In sub-letting any work all contracts with his sub-contractors must be made under these general instructions and conditions.

The names of sub-contractors must be submitted to the Commissioners and Architect for their approval, and in no case will any work or material be accepted that may be furnished or executed by sub-contractors not approved of.

The contractor shall be held fully responsible for all contracts with his sub-contractors as to the furnishing of materials and the performance of labor by the sub-contractors of any part of the work.

Neither the Commissioners or Architect will be party or parties to such contracts, and will not give any decision or opinion in case of rights, privileges or disputes between the contractor and sub-contractor.

The contractor shall be held fully responsible for the work of his sub-contractors of the various departments of the work.

The contractor shall assume and become liable for all losses or damages by fire or otherwise to the work or materials either in the building or on the premises, until all work on his contract is fully completed and accepted, and shall insure the same against loss by fire in such companies and to such amounts as may be satisfactory to the Commissioners, the policies to be made payable to the county and the contractor as their respective interests may appear, the amount to be increased from time to time as the work progresses. The contractor must also pay the additional premium for builder's risk.

Immediately after the contract is entered into, the contractor shall proceed with the fulfillment of its requirements and prosecute the work with the greatest reasonable dispatch in such a manner and by such methods as to bring the work to completion on or before the stipulated time.

The contractor shall have such appliances and materials on the premises as will satisfy the Commissioners and Architect that he is not only willing, but able to comply with the requirements in a practical and ready manner.

The different departments of the work and every detail thereof shall be carried on in such order of procedure and at such times and seasons, as will in no wise hinder or interfere with the progress of any department, detail thereof or the work of other contractors or sub-contractors, who shall have equal rights on the premises for the performance of their work.

The Commissioners and Architect reserve the right at all times to decide whether or not the contractor is faithfully and vigorously prosecuting the work either as a whole or in detail.

Before commencing other work, the contractor shall carefully and thoroughly examine all work that has been commenced or finished that may be liable to affect the work to follow.

If any previous work has not been done properly, the contractor shall suspend, or not commence any other work until such defective work is corrected and made good.

No excuse will be entertained for any portion of the work not being right because other work is wrong.

The contractor shall verify all measurements, and will be held responsible for the correct laying out of the work, and in the event of any error he shall do what is necessary to remedy the same.

The contractor shall furnish, at his own cost and expense, and without extra charge, all transportation, mechanical appliances, power, labor and materials, necessary to properly prosecute the work to full completion in every department and detail in the most satisfactory manner.

All machinery and appliances of every kind used by the contractor upon or about the work shall be deemed in the possession of the Commissioners, to be held as security for the full performance of the contract in addition to the security by bond given by the contractor, but such appliances and machinery shall be cared for by the contractor, and he shall be responsible for any loss or damage occurring to them, and said machinery and appliances to be delivered to the contractor only when he has fully completed the work for which they were intended.

The contractor shall provide all incidental work, materials and appliances necessary to the proper and entire finishing of the work.

The contractor shall at all times employ a sufficient number of skilled mechanics and artisans for the proper performance and execution in the best and most thorough workmanlike manner of every department of the work.

The contractor shall distinctly understand that each and every department of the work and detail thereof must be perfect and complete in itself, and that all material, apparatus and appurtenances of every kind and description shall be of the very best of their respective kinds, and that all work shall be done strictly in the most thorough workmanlike manner that the best skilled mechanics and artisans can execute, and it is therefore required of him that he use due diligence to inform himself of everything connected with the entire work.

All work and materials of whatever description, not in strict accordance with the requirements of the drawings and specifications, shall be rejected by the Architect, and the contractor shall, within three working days after receiving either verbal or written notice from the Architect, remove from the grounds all material, whether worked or unworked, and shall take down all portions of the work that may be condemned as unsound, improper, or in any way failing to conform to the requirements of the aforesaid drawings and specifications, and in case of failure so to do, the work shall cease until the same is made proper and right.

Should any dispute arise as to the character or quality of the work executed or the materials furnished, or as to the way and manner of the execution of any part of the work performed, a decision in each and all cases shall be based only on the requirements that the work and materials shall be in every respect the very best of their respective kinds as herein provided, and shall fully comply with the plans for said building and these specifications, and that which may be considered as customary or usual shall in no wise enter into any consideration or decision whatever.

All work and materials shall be subject at all times to inspection by the Commissioners and Architect, or their representatives, and the contractor shall provide proper facilities for such inspection as may be required by them.

The inspection of work and materials shall in no case be construed as an acceptance of the same, or to relieve the contractor from any of his duties or obligations herein set forth or required in the contract.

It must be distinctly understood and agreed that under no circumstances whatever will the contractor be permitted to make any alterations or deviations in the execution of the work or in the materials to be furnished, different in any respect from what is required by the drawings and specifications, without the written consent of the Commissioners and Architect. And the contractor shall make no claim for extra work of any kind unless the same was ordered in writing by the Commissioners and Architect.

The contractor shall be held fully responsible for the safety and good condition of all work and materials embraced in his contract, and under his charge, until the completion of his contract as an entirety.

The contractor shall clear away from time to time all debris and waste materials, resulting from his operations, so as to keep the premises in a clean and orderly condition.

The contractor shall be held responsible for using improper materials or workmanship, and for all damages arising therefrom, even though his attention should not be called to it. He shall amend and make good, at his own cost and expense, any defects, settlements, shrinkage or other faults in his work arising from defective or improper materials or workmanship which may appear during the progress of the work, or within twelve months after the completion of the work, and should any damage occur to any part of the work or materials, the work and materials damaged shall be removed and perfect work and material replaced to the entire satisfaction of the Commissioners and Architect.

Any work or material damaged by any alterations to be made good at the contractor's expense.

The Commissioners reserve the right and power, at any time, to add to or deduct from, or to make such alterations, deviations or changes, either in the construction, detail or execution of the work or materials as to more fully or completely fulfill the general plan and design of said work that they may deem necessary or desirable, without in either case invalidating or rendering void the contract or the bond securing the faithful performance of the same or impairing them or either of them in any way, shape or manner, but said contract and bond shall remain in full force and effect the same as though such changes, alterations, deductions or additions had been made before said contract or bond had been executed in the first instance. But in case the contractor and the board are unable to agree upon the price to be added or deducted on account of such changes or alterations, the same shall be fixed by the Architect, and if either the contractor or the board is dissatisfied with the price so fixed by the Architect, the same shall be referred to three arbitrators, one to be selected by the board, one by the contractor, and the other by the two so selected, and the decision of any two of such arbitrators as to such price shall be final and conclusive.

In case such changes may be required during the construction of said work, the same must be consented to by the contractor and carried into effect.

The Commissioners have appointed Brentwood S. Tolan, Architect and Supervisor of the entire work, and empowered him with authority to enforce all the conditions of the contract, specifications and drawings, and to give such information, either in language, writing or drawing, as in his judgment the nature of the work may require; to inspect and approve or reject any or all material and work, having particular care that any and all work done, and material used, be such as is required and described herein; to make all estimates and give all certificates that the contractor may be entitled to, of the amount due him on the contract price for material and work; determining the amount of damages that may accrue from any cause under the contract and these specifications, and settling all deductions from, or additions to contract price growing out of alterations or changes of the designs and plans after the same is under contract, and to furnish all necessary drawings and information that may be required to fully illustrate the designs given, but all such inspection, rejection, approval and determination of damages shall also be subject to the rejection or approval of the Board of Commissioners to whose satisfaction all of said work shall be done, and subject also to the right of reference to arbitrators as hereinbefore provided as to the price to be allowed for alterations. And when the work is fully completed and finished in all its parts and details in strict compliance with these specifications and the said plans and drawings, to the entire satisfaction of the Commissioners and Architect, then to issue a certificate of acceptance to the contractor, which certificate, if unconditional and approved by the Commissioners, shall be an acceptance of the work, but shall not relieve the contractor from any liability on account of any imperfect or defective work, or from any responsibility on account of any guarantee for said work or any part thereof.

The Commissioners have appointed William H. Goshorn as Engineer and Superintendent of the work, who is empowed with authority to inspect and approve or reject any or all material and work under contract, and to give all directions and orders to the contractor that he may deem necessary to the advancement and welfare of all the work, subject, however, to the approval of the Board of Commissioners in all cases, as hereinbefore provided in case of the Architect.

All departments of the work are to be erected and finished agreeably and in conformity with the true intent and meaning of the specifications and drawings now on file in the County Auditor's office.

The drawings consist of plans, elevations, sections and details, showing the general disposition, construction and character of the work.

The present detail drawings are intended to explain the general character and details of the work to be executed.

The drawings will be further supplemented, during the progress of the work by large scale and full size details.

The drawings and specifications are intended to co-operate and to provide for and comprise everything necessary to the proper and complete finishing of the work; but any work shown on the drawings and not particularly described in the specifications, or called for in the specifications but omitted on the drawings, and any work reasonably implied and evidently necessary for the complete finishing of the work, although omitted from both, is to be done by the contractor without extra charge, the same as if it were both shown and specified, unless such omissions are expressly mentioned in these specifications.

Such corrections are not to effect any contract that may be in force, and no advantage to be taken of any manifestly unintentional omissions or discrepancies.

As evidence that said examinations, calculations and comparisons have been made, the signature of the contractor will be required upon each and every drawing, and the specifications on file in the Auditor's office.

The Architect will give full written and verbal explanation to the contractor as to any point which may be obscurely expressed in the drawings or specifications, or not clearly understood by the contractor.

Should it appear that any part of the work is not sufficiently detailed or explained to enable the contractor to thoroughly understand the nature, character and quality of the work required, and all matters relative thereto, the contractor shall apply to the Architect for such further drawings and specifications as may be required, and if he deems the request reasonable, they will be furnished.

The contractor shall give ample notice, so that they may be prepared by the time they will be needed, and such additional drawings and specifications shall be considered as part of the original drawings and specifications, and shall not operate to extend the stipulated time for the completion of the work or otherwise relieve the contractor from any responsibility under the contract.

In the event of any doubt or question arising regarding the true intent and meaning of the drawings and specifications, reference shall be had to the Architect, whose decision shall be final and conclusive.

The contractor shall furnish such dimensions, drawings and specifications of all machinery and apparatus as the Architect may from time to time call for. He shall also furnish the Architect with duplicates of all drawings and specifications furnished by him or his sub-contractors and intended for use in any portion of the work.

In case any work is to be executed according to dimensions, drawings and specifications furnished by the contractor, or his sub-contractors, they must be submitted to the Architect for approval, such approval will be only as to general detail, and the contractor shall be responsible for any incorrectness. No drawings or specifications other than those furnished by the Architect, or examined or approved by him, will be permitted to be used for the execution of any portion of the work or any detail thereof.

The contractor shall have copies of all drawings and specifications conveniently accessible about the work for the use of all persons concerned.

All drawings and specifications are instruments of service to be used for this work only, to remain the property of the Architect, and returned to him on completion of the work.

The copying of any drawing will not be permitted except by the consent of the Architect. All duplicates will be furnished by him that may be necessary.

Should the contractor object to any decision of the Architect, he has the right to appeal to the Commissioners, and all appeals or statements to them shall be made in writing, setting forth the facts in particular, and the decision or statement from the Commissioners will be rendered in writing, and shall be considered binding by the contractor. The decision of the Commissioners shall be considered as final in all questions that may be referred to them.

Any decision of the Architect shall not be final as against the Commissioners, as they may require any change of the same to fully conform to the contract.

In case of any unusual or unnecessary delay or inability of the contractor in providing or delivering the necessary materials, or performing the necessary work at the time required, then in such case the Commissioners within five days after having notified the contractor of their intention so to do, shall have the right, under the contract to enter upon the work; and after giving ten days notice by publication in a newspaper printed and published in Fort Wayne, Indiana, re-let such necessary labor and materials to be performed and furnished, or may without notice, purchase the necessary material and employ the necessary labor, as the case may be, and carry the work on to completion in such manner as in their opinion shall be proper and right, charging the cost thereof to the contractor and deducting such costs from the contract price; or, if at any time the contractor shall fail to comply promptly with the notice of the Commissioners to prosecute the work in effect, or cease entirely the work in effect; or, if the Commissioners shall be of the opinion that the contractor has abandoned or intends to abandon the work, then in either case the Commissioners shall have the right to enter and take posession of the work and all materials, machinery and implements of whatever kind belonging to, or in use by said contractor, and upon notice aforesaid to re-let, or without notice, employ men and purchase the material necessary to complete the work; and if in any case such work shall cost more than the balance of the price agreed to be paid for doing the whole work over and above what had already been paid to the contractor, such extra cost and ten per cent. added thereto shall be charged to the contractor and may be retained from any moneys that may be due on the contract; or, if there is not that amount due, the same shall stand as liquidated and agreed damages to be paid to said board by the contractor and his bondsmen executing the bond given to secure the performance of his contract. The contractor will be held responsible for all defects that may arise from bad workmanship or improper materials, and for all damages to the work that may occur or be sustained on account of the doing of such work after he has quit or abandoned the same.

Estimates will be made monthly during the progress of the work, but not within twenty-five working days from the time of the previous estimate.

All estimates will be made to the contractor, by the Architect, for labor performed and material furnished and delivered on the premises, but no estimate will be made for either that is not in strict accordance with the requirements of the drawings and specifications, or have not become permanent parts of the work, or delivered on the premises for that purpose.

Ten per cent. of the amount of each estimate will be retained for the faithful performance of the contract, until the acceptance of the work by the Commissioners.

The contractor shall notify the Commissioners and Architect in writing three days or more before the day on which he desires an estimate, but any oversight, failure or refusal of the Architect to furnish such estimate shall not authorize the contractor to annul or abandon the contract, or cease doing said work, but in either such case the contractor shall apply to the Commissioners for such estimate.

On the day previous to making the estimate, the contractor shall prepare a written itemized and detailed statement, giving the kinds and quantities of material, prices thereof, amount of labor, etc., for which he desires an estimate. Said statement shall be full and complete in all respects, and shall be certified to as correct by the contractor.

Said statement shall be presented to the Architect as early as possible, and should it be approved by him, he will issue a certificate on the Commissioners for the amount due the contractor on said estimate to be made by him, and if approved by the Commissioners, they shall within five days thereafter issue to such contractor an order on the Treasurer of said county for the amount of such estimate less ten per cent. of the amount thereof to be retained until the final completion and acceptance of the work, as herein provided.

All estimates will be based upon the actual amount of material furnished and labor performed, and no estimate will be made otherwise.

The contractor shall, at the time of making the second

estimate, and each subsequent estimate, submit to the Commissioners satisfactory vouchers and receipts, showing that he has paid for all materials and labor included in the previous estimates, and at each estimate and payment to the contractor, he shall receipt for the same, and in said receipt shall state that he has no further claim for work or material and no further claim to date other than the retained ten per cent.

All payments are to be on account of the contract, and in no case are to be considered as an acceptance of the same, or of any work or materials until the final estimate and certificate of acceptance is approved by the Commissioners.

Upon all estimates being made, such material and artisan work as shall have been approved and accepted shall be considered as the property of the county, and when any materials have been delivered on the premises, and accepted as suitable, although no payments have been made upon them, such materials shall be considered as in the possession of the Commissioners, and shall be held by them as security for all payments made and all liability made upon the contract.

Should the contractor fail to do and perform all and every one of the obligations required in the drawings and specifications, or the articles of agreement, or fail to complete entire his contract, then in such case the retained ten per cent. shall be forfeited to said county, in addition to the liability on his bond, as part of the damages which said county may incur by reason of the failure of the contractor to complete his contract.

Should any work or materials be found, after payment of the same, not to be as required herein or on the drawings set forth, said work and materials shall be removed, and the sum paid for the same shall be deducted from the following estimate after rejection, and when the work and materials have been satisfactorily replaced and made good, then the sum deducted shall be paid the contractor with the following estimate.

Should the work be unneccessarily delayed or any delay in the stipulated time of final completion occur, the cost of extra inspection and supervision necessitated thereby shall be determined by the Commissioners and borne by the contractor.

The entire work must be delivered over to the Commissioners in a clean and undamaged condition fully cumpleted and in full accordance with the requirements hereof, and the plans and detail drawings therefor, on or before the first day of August, 1899, and the contractor shall pay to the Board of Commissioners the sum of fifty (50) dollars per day for each and every day thereafter that said building is not delivered and completed as aforesaid as agreed and liquidated damages.

When all of said work is fully completed in all its various parts and details, in strict accordance with the requirements hereof, and with the true spirit, intent and meaning of the aforesaid drawings, specifications and the contract, and to the entire satisfaction of the Commissioners and Architect, then in such case the Commissioners and Architect will issue to the contractor a final estimate and certificate, showing the completion and acceptance of the work, and thereupon the Commissioners shall execute to the contractor a warrant upon the Treasurer of the county for the full amount then due the contractor.

It will be required of the contractor to have vouchers and receipts in full from all parties who have contracted with him for labor and material for said work.

The contractor shall notify the Commissioners and Architect that he is ready to have a settlement, so that if they have any statement to make they can do so before the Architect makes his final estimate of adjustment between all parties concerned.

The Commissioners reserve the right at any time to let and contract with other contractors for any work or material outside of the original contract, and the contractor shall consent to the same and permit such extra work to be executed, and permit all parties connected with such extra work and contracts to have access to the premises.

It is intended that these specifications together with the drawings herein referred to, shall cover complete and entire each and every department, except what is specifically omitted, and the bidder is requested, for the purpose of furthering his interests, as well as to facilitate the work, to carefully examine the same, and in case of discovery of any omissions which would effect the perfect completion of each and every department and detail thereof, to consider such omission or omissions as fully corrected before making his bid, as complete and perfect work in every detail will be expected and demanded of the contractor, except as to what is specifically omitted.

It must be understood and by the acceptance of these specifications is agreed, that all work is to be executed according to the true intent and meaning of these specifications, and the drawings herein mentioned, and to the satisfaction of said Board of Commissioners, and that any labor, material. apparatus or appliances not herein specified or shown on the aforesaid drawings, but which may fairly be implied as essential or proper to the thorough erection and completion of any and all parts of the work and details thereof as proposed, shall be furnished and installed by the contractor without extra charge, notwithstanding each and every item necessarily involved in the work is not specifically mentioned herein or shown on the drawings.

All contracts drawn up between the contractor and his sub-contractors, shall state specifically what portion of the work in particular are included, and what is not included in the work between them. Where the different branches of work are directly or indirectly connected with each other, the line of separation must be drawn by the Contractor and his sub-contractors.

The foregoing conditions and requirements are intended and shall govern each and every department and detail thereof in the true meaning and intent of the following specifications:

OLD COURT HOUSE BUILDING.

The Commissioners will have the old Court House torn down, and remove the old material from the site, including the present stone pavement and cement curbing.

They have contracted with Edward A. K. Hackett and Sidney C. Lumbard, for the erection and maintainance of a closed fence around the premises, about eight feet high, and have provided in said contract for certain gates and sections of the fence to be used by the contractor in constructing the building, which contract is on file in the Auditor's office, and spread at length upon the Commissioners' record. All bidders must acquaint themselves with all the provisions of said contract, and comply therewith. The contractor must use all reasonable diligence not to damage or injure said fence, and shall keep all gates closed when not in use.

The contractor shall take the premises in the condition it is left after the old building, pavements, etc., have been removed. He shall also provide and maintain on the Court House premises an office for the use of the Commissioners, Architect and Superintendent.

SURVEYING.

The contractor shall do all surveying. Accurate lines of the work to be established and permanent points maintained for guidance.

EXCAVATIONS.

Before commencing the excavations, the contractor shall remove all debris, waste materials and all other obstructions from the premises, so as to leave the site of the buildings clear.

Excavate the areas sufficiently larger than the buildings require to allow for convenient working, and to the depths and levels shown on drawings.

Excavate for all footing trenches, tunnel, sewers, drains, water and gas pipes, and all other requisite works as indicated on drawings and herein described. Test pits to be sunk so as to determine the exact nature of the soil.

The bottoms of all footing trenches to be level and the sides plumb.

Excavate trenches for steam runway between Power Station and Jail, also for natural gas pipe between street curb and boiler room.

The excavations must be kept thoroughly drained, and dry as possible at all times.

Shore the sides of the excavations where necessary in a substantial manner with plank and heavy inclined shores wedged to solid bearings. .

As the footings and foundations progress, the excavations are to be refilled and graded up to within one foot of the finished grade, floor and pavement surfaces.

Fill in around all walls, and drains, with good sand and gravel.

All filling to be thoroughly rammed and made compact

as the work progresses, so that it will not settle after the work is completed.

Should there not be filling enough the contractor is to supply the deficiency; or, if there should be a surplus of excavated material, the surplus, together with all debris and other waste material to be removed from the premises, to the land lying North of the Jail, on North Calhoun Street, in said city, and at such points thereon as may be designated by said Commissioners.

DRAINS.

Lay around at the footings of all outside walls best quality six-inch drain tile, to have sufficient fall and proper connections with sewer. Connect a blow-off drain from boilers to drain tile entering sewer. An approved trap to be connected with this tile in the boiler room floor.

A four inch drain tile to be around under all floors of the Power Station, and connected with blow-off drain from boilers.

The Tunnel drains to be connected with this work.

The down spouts of both buildings to be connected with glazed tile socket jointed pipes, best quality, none to be less in size than the down spouts, and to increase in size with every connection. All to have the necessary branches, elbows, knuckles, traps, vent pipes, etc., complete.

Bed all pipes carefully, and make joints tight in Portland cement, pointing outside and cleaning off inside at laying of each length. To be connected with sewer in the best manner.

CONCRETE FOOTINGS.

Before commencing the concrete footings, the contractor shall examine, test and be responsible for the firmness and condition of the soil. He shall do all borings and soundings at all points, and to the depths that may be necessary to make the required tests and examinations.

All foundations for walls, piers, stack, tunnel, boilers and machinery, to be started on concrete footings, excepting the tower piers which will be started on stone footings, the stone footings to be bedded on footings of steel beams imbedded in concrete as hereinafter specified.

All concrete for foundation work is to be composed of one part best Portland cement, three parts sand, and five parts crushed stone.

All sand to be clean, sharp and free from loam or dust.

All crushed stone must not exceed one and one-half inches in size; and must be good, hard lime stone. All to be kept perfectly clean and free from dust.

Before depositing the concrete, the trenches must be well cleaned out, and their entire areas thoroughly tamped with heavy mauls. Saturate the trenches, when necessary, with grouting of sand and cement to retain the moisture of the concrete until set.

The sides of all trenches and footings to be curbed with plank. The concrete to be dumped into the trenches between curbs. Saw all planks to form true angles and required lengths. They may be removed and used again, but must remain until the concrete has become thoroughly set.

The concrete must be made near the trenches in which it is to be deposited.

The cement and sand shall be thoroughly mixed dry in boxes, then with a rose head wet with a small quantity of water, and temper to the consistency of a stiff mortar, after which it shall be spread evenly in the boxes, and the proper amount of clean wet broken stone added, and the whole thoroughly mixed, laid in the trenches immediately, in layers not more than four inches thick, and tamped perfectly solid until the water rises to the surface. Fill in fully the widths of trenches, and of such depths and dimensions and number of courses as required by the drawings, or as may be directed. Every layer and course being allowed to set before the succeeding ones are placed, and the joints so lapped that the concrete be one continuous footing. Every layer to be carefully and thoroughly tamped to position, level and of even thickness, making a thoroughly compact mass.

The last course must be properly prepared for receiving the work to follow. All concrete to be thoroughly protected from injury, and to remain undisturbed throughout.

The concrete to be kept thoroughly clean and damp until after the full thickness of any portion of it is completed, when the surface will be covered with sand, saturated with water.

The concrete to be thus kept covered and the sand wet until the concrete has become thoroughly indurated and the work to follow ready to be commenced, when the sand will be removed.

After each layer has been allowed to set, it shall be swept clean, wetted and made rough by means of a pick before putting on the next layer.

Should, at any points, the arrangements of pipes or drains necessitate their passing through the concrete, terra cotta pipes of large size are to be imbedded in the concrete so as to leave a permanent opening.

There will be concrete footings for columns supporting work in West Vestibule, and the partitions of Commissioners' Court Room and ceiling above, also under East Vestibule, noted on Foundation Plan.

STONE FOOTINGS.

There will be stone footings under the tower piers, of the thickness and dimensions shown on drawings. Under each pier the stone will be one piece, perfectly squared, and to be of uniform thickness throughout. All beds to be level and true, to be scappled if necessary to make them so. All to be of good, sound and best quality flat bedded lime stone from either the Greensburg or St. Paul quarries, to be approved by the Commissioners and Architect.

Extra care will be required in quarrying to select stone that has level and flat beds, and quarried to make square sides. All dressing of the stone must be done before they are set in place, and thoroughly cleaned.

In setting these footings it will be required that they be lifted and set with derricks. Under no circumstances will boards, rollers or bars be allowed for the purpose. All must be lifted clear and set exactly in their places.

The stone footings will be bedded on the concrete and steel footings in a thick floating coat of cement mortar, and thoroughly bedded with a bedding maul in weight requiring the strength of two strong men to handle. All strokes to be in the center of the stone, no strokes on ends or sides; all must be bedded to line on edges and top, and made perfectly level.

STEEL FOOTINGS.

Steel beams, of the number, size and weight shown on foundation plan and sections, will be imbedded in concrete. forming footings for the tower piers. Strong, tight boxes

are to be used for the concrete work, and all space between
the beams and boxes to be completely filled with concrete,
to the required levels, forming a compact mass. This concrete will be in the form of cement grouting. No stone will
be used, as the beams at their flanges will be only about one
quarter of an inch apart. All beams to be painted before
being placed in position; the painting to be as specified for
structural steel and iron work. The cement to be thoroughly
set and indurated before the stone footings are placed upon
this work. All to be thoroughly protected from injury.

DAMP PROOFING.

The outside surface of all foundation walls of Court House
and Power Plant to be plastered between the footings and
stone base course at grade lines, with cement mortar, three-
quarters of an inch thick, composed of one part cement to
two parts sand; to be evenly applied, troweled smooth and
thoroughly protected from injury.

BRICK WORK.

All foundations, walls and piers above the concrete footings to be of brick work, including stone backing and other
work denoted on drawings, and of the dimensions figured
thereon.

All brick used throughout the entire work must be the
best quality, well tempered, hard burned, square edged, uniform in size and texture, free from lime stone and large peb-
bles, and to have clear, sharp ring. All of the best selection
from the kiln, and to be approved by the Commissioners.
The brick must be out of the kiln at least ten days before
they are used in the work.

No bats, swelled, over-burned, or refuse brick will be
allowed; no salmon, dry-pressed, soft or defective brick will
be allowed at any point in the construction, or at any time on
the premises.

All brick to be laid with shore joint.

All brick to be laid wet in dry, warm weather, and dry in
damp or cold weather. No brick to be laid in freezing
weather, except when absolutely necessary, in which case the
brick are to be heated and the mortar is to be hot, and to
contain enough salt to prevent freezing. Such brick work
is not to be pointed up until Spring.

The entire thickness of all walls and piers must be built up even, and the courses of brick work to be kept perfectly level throughout, and all built uniformly, plumb and true to line on all sides.

No portion of the work to be built up more than a scaffold high before the remainder is brought to the same level.

All work to be of the dimensions shown on the drawings, to be brought to the exact heights and the offsets, to bring the work to the required thickness, to be made at the tops of the beams. The tops of all walls and piers to be carried their full dimensions between all iron work to the top of same, and built up under the roof between the iron work.

All brick work must be well laid and thoroughly bedded and bonded in mortar, the brick laid as close to each other as possible. The joints of each and every course to be flushed full with mortar, well worked in with the trowel, leaving no voids. Throughout the entire work the brick work shall be bonded with header binders, every fifth course through the whole thickness of the walls, piers and stone backing.

All walls that are to be plastered to have joints left rough; all walls to be left unplastered are to have joints struck smooth and pointed up.

All chases, channels, flues and openings for the ventilation, heating, plumbing and electric wires to be accurately laid out to conform to measurements as shown, and as may be directed and required. All to be done as the work progresses.

There will be air inlets under all outside windows above the Ground Story. To be carefully and accurately constructed according to the drawings, and as will be directed. The brick work over the openings to be supported by wrought iron lintels and wherever arches may be used.

All flues to have struck joints and kept clear of all obstructions. Special care must be taken with the flues that are to be run out of the perpendicular, that they are kept perfectly clear of all obstructions. Flues to be plastered wherever directed.

Turn strong brick arches over all brick openings that are not provided with stone or iron lintels, or wherever required, either semi-circular, segmental or flat. Wherever flat arches are used, segmental relieving arches are to be sprung over them, as the nature of the work may require.

The large arches in the tower must be constructed in the most careful manner, to be perfectly bonded the full thickness of the wall with alternate headers and stretchers.

All arches to be bonded clear through with as many rowlocks as their importance requires, and as directed by the Architect.

Build rowlock heads to inlets and outlets of all flues wherever required.

All centres to be accurately and substantially constructed, and the arches to be left perfect on removing the same.

The large arches in the Power Station will require special attention.

The contractor shall build in all iron anchors, tie rods, and all other iron work entering brick work, properly bed all plates, beams, lintels, etc., true and level. All bolts for plates, etc., not otherwise specified, to be placed with templets. All to be thoroughly and securely imbedded in mortar and bricked up solid as the work progresses.

All heights of courses to accurately conform in levels to that determined at beginning or at different stages of the work, that the stone and brick may be well and thoroughly bonded together.

Back up all stone work so that the core of the wall will be as good and as solid as the faces, by using flat tile or slate to level up with.

The brick work must follow the stone work closely, backing up the stone immediately after it is set, the brick work well leveled up tight and solid to the same, in a condition to receive the next course of stone.

Build all interior work in proper manner, so as not to delay the setting of the stone. Special care will be required in the bonding of brick and stone work, and leveling up of same full and solid, filling all voids.

The brick, stone and iron work must be carried on in such manner that neither will interfere with the progress or stability of the others. No toothing will be allowed in any event.

The brick backing to the terra cotta work around the dome and lantern of tower must be anchored to the steel frame work with wrought iron anchors as directed. The terra cotta work of the Power Station will also be properly backed up and anchored as required and directed.

Double rowlock brick arches to be turned between steel beams, forming driveways over the coal areas on the North and South sides of Power Station. These brick to be hard, and laid in cement mortar as specified. Build in cast iron rings for coal chutes as shown.

The foundation walls for the boilers are to be built up within one foot of the surface of the boiler room floor, ready for the boiler setting.

All outside walls of the Power Station, from the stone base course at grade line, to be faced with selected face brick, of uniform size and color. To be all stretchers well bonded to the wall every fifth course by a row of headers laid diagonally into the backing. All joints to be not more than one-quarter inch thick, smooth, concave, pointed, and a plumb bond maintained throughout. The mortar for this work to be made with lime putty and white sand or marble dust. All to be well cleaned down on completion with a weak solution of muriatic acid.

All brick work below the top level of the Court House and Power Station base courses above guard lines, shall be laid in mortar composed of one part Portland cement, to three parts clean sharp sand, mixed dry in a box, afterwards wet, tempered and immediately used.

Above these levels the entire brick work throughout said buildings, with the exception of the tower piers and the piers at each side of the four main entrances, shall be laid in mortar composed of one part Portland cement to three parts lime mortar, the lime mortar shall be composed of best quality fresh burned lime and clean sharp sand, all properly tempered and made in the very best manner. The lime to be well slacked and run through a fine mesh, and all sand to be screened.

The twelve piers of the tower and the eight piers at the sides of the four main entrances to the Court House, to be run up to their extreme heights with mortar composed of one part Portland cement, to three parts clean sharp sand, same as specified for foundation work.

All cement must be of the best quality of Portland cement, and shall be subject to inspection. The Architect reserves the right to take samples for testing from every lot of cement delivered. The contractor shall notify the Architect at an early date the brand of cement he wishes to use, in order that the proper tests may be made by him.

All rejected cement shall be at once removed from the premises by the contractor. He shall provide a suitable place for storing cement, where it will be thoroughly protected from the weather. He shall have a sufficient quantity stored in advance to allow making the seven day and twenty-eight day tests, without interfering with the progress of the work.

Ten per cent. by weight of the cement shall pass a No. 100 sieve, 10,000 meshes per square inch.

Two cakes shall be made of neat cement, two or three inches in diameter, one-half inch thick, with thin edges. It shall take at least thirty minutes for these cakes, when mixed with water having a temperature between sixty and seventy degress Fahr. to the consistancy of a stiff plastic mortar, to set hard enough to stand the wire test. one-twelfth diameter loaded to one-quarter pound. and one-twenty-fourth wire loaded to one pound.

One of these cakes, when immersed in water, after it has entirely set, shall show no signs of cracking or disintegrating.

The color shall be uniform throughout.

Briquettes shall stand the following strains without breaking:

Neat cement seven days, one day in air, six days in water, four hundred pounds.

Neat cement twenty-eight days. one day in air, twenty-seven days in water, six hundred pounds.

One part cement to three parts sand, by weight, seven days, one day in air, six days in water, one hundred and twenty-five pounds.

One part cement to three parts sand, by weight, twenty-eight days, one day in air, twenty-seven days in water, two hundred pounds.

The tests shall be made by an engineer selected by the Architect, and shall conform in all respects to the recommendations of the committee of the American Society of Civil Engineers on a "Uniform System for Tests of Cements."

All expense incurred in making tests shall be borne by the contractor.

Back of the large stone arches outside of Court Room No. One and Four, there shall be applied on the brick and stone backing a course of best Seyssell natural asphalt damp proofing, to cover the inside surface one foot more all around than the size of the arch outside, and one foot below the window sill line.

There will be a brick trench, cement lined, ten inches wide by sixteen inches deep, for the natural gas pipe, to run from the South side of Power Station to front of boilers. This trench to be provided by the contractor.

For brick work of Tunnel and Stack, and fire brick for Stack and Boilers, see specifications under their respective headings.

The Entrance Vestibule, Machinery Hall and Boiler Room of the Power Station, to be lined with enameled brick from their respective floor levels to the ceiling cornices. These bricks are to be thoroughly enameled, not glazed, and equal in quality and surface to Tiffany Co.'s best enameled brick. No dry pressed brick will be accepted. All of this work to be thoroughly secured to the walls and laid in the very best manner. All joints to be neatly pointed; none to be over one-eighth of an inch thick. Three different colors of enameled brick will be used, and these, together with the color of the joints, will be subject to the approval of the Architect. This work will form bases, wainscoting, work around doors, windows, arches, etc., as shown on drawings.

GRANITE.

The plain base course at grade line, around the entire Court House, all entrance steps and platforms, the jambs and reveals, imposts, pilasters and columns at the four entrances, the plinths, bases, shafts and capitals of said imposts, pilasters and columns, will be of North Jay granite, from the North Jay, Maine, quarries, or any other granite acceptable to the Commissioners, of a color to harmonize with the Blue Bedford lime stone of the superstructure. A darker color than the stone is preferred for the base course only. All to be best selected No. 1 stock, close grained, of good texture, free from all defects, and to be of even, uniform color.

Samples of granite to be submitted for approval; their adoption depending on the decision of the Commissioners. Bidders will state in their proposals the granites they offer.

The base course to be bedded level at least four inches below grades, and stepped to conform to the grades shown on drawing of grades, etc., the highest point being at the Court Street entrance, the Ground Floor level being six inches above this point.

The base course will be level on top all around, corresponding to grade in height, and have a minimum thickness of one foot and six inches, the heads at return angles to be not less than two feet and two inches wide. The backs to be pointed off and plumb.

The beds, builds and joints are to be cut fair, true and out of wind, full to the square and have an equal bearing throughout. All arises sharp and all side joints verticle, extending to the back, to be so worked that the joints will not exceed three-sixteenths of an inch in thickness.

The face of the base course will project two inches from the Blue Bedford lime stone moulded base above.

The wash of the base course will be two inches by one inch, to be tooled eight-cut work, and the finish of the face to be fine axed. All steps and platforms to be ten-cut work. The jambs and reveals, all plinths, and the shafts of imposts and pilasters to be ten-cut work, all bases and capitals smoothly rubbed, and the shafts of all columns to be highly polished. All capitals to be carved as shown on detail drawings. The base course to be bedded and pointed in cement mortar as specified for foundation work, and the work above base course to be bedded and pointed in mortar specified for lime stone work. All door sills to run back to the center of the thresholds.

BLUE BEDFORD LIME STONE WORK.

The Board of Commissioners have selected and adopted for the building of said Court House, the very best quality of Selected No. 1 Blue Bedford stone, taken from the "Hoosier Quarry," now owned and operated by The Bedford Quarries Company, which stone shall be cut and taken from the ledge or depth in said quarry as selected and designated by said Board of Commissioners, or its agent or representative, and must be perfectly sound, free from veins, seams, holes,

crowfeet, iron, spots, streaks, lines, oil stains and all other
staining properties and defects of every kind and description, and shall and must be absolutely uniform in color and
texture throughout the whole, and within the terms and
specifications and requirements herein mentioned, to the
satisfaction of the said Board of Commissioners, and the
Architect and Superintendent of said building. and shall be
subject to inspection, rejection and approval by said board
and the Architect and Superintendent of said building, at
Fort Wayne, Indiana, and even after the same is placed in
said building.

All stone conforming to these requirements shall be cut,
dressed, carved, sculptured, and finished by the contractor
within the corporate limits of the City of Fort Wayne, Indiana.

The Bedford Quarries Company owning and operating
said "Hoosier Quarry," has executed a written contract and
bond, which are now on file in the Auditor's office, by which
it agrees to furnish and deliver the character and quality
of stone hereinbefore specified, and in the quantities required
for building said Court House, to the contractor. F. O. B.
cars at Fort Wayne, Indiana, at a price not to exceed thirty-
seven cents per cubic foot. All bidders are requested to
fully acquaint themselves with all the provisions of said contract and bond before submitting their bids.

The contractor, in order to protect his own interests, will
notify the Commissioners and the Architect and Superintendent of every stone that does not conform to the requirements, before cutting or placing any in the building that may
be rejected on account of not fully conforming to the requirements.

But such notice and any inspection made by such Commissioners or Architect and Superintendent shall not restrict
nor limit the right of such Commissioners or Architect and
Superintendent to reject any stone not meeting the above
requirements, even after the stone may have been placed in
said building.

The contractor will be held responsible for all expense
that may be incurred in cutting, carving, sculpturing, finishing and placing any stone in the building that may be
rejected, and the expense of taking out the rejected stone,
and the expense of replacing and making good all material

and work, which he shall be required to do on account of such rejection.

Should any of the stone become injured in any way or manner, before or after they are set, they will not be allowed to be used or remain in the building.

All stone work of the exterior of the Court House, from the top of the granite base course at grade line to the springing of the main dome of the tower, shall be of approved Blue Bedford lime stone, as herein specified, excepting the jambs, imposts, pilasters and columns at the four entrances, which shall be of granite as before specified.

All the cut stone work must be of the dimensions, forms, designs, profiles and jointed, as indicated on the drawings, and in all cases to strictly conform to details to be furnished by the Architect.

The beds, builds and joints must be full, true and out of wind, and to have an equal bearing throughout, to be so worked that when the stones are set, the joints will not exceed three-sixteenths of an inch in thickness.

Figured dimensions on drawings include the joints.

No holes or depressions will be allowed in the beds, and in no case will beds projecting back from the surface be allowed. All to be cut full to the square and all abutting joints to be fine pointed, square to the beds and faces. All to have true angles, joints and arrises.

The beds of columns, bases and capitals must be worked perfectly true and rubbed for close fine joints.

All work to be straight, true, level and plumb, and all joints and beds, except sills, close and full. Slip sills to be bedded their entire length; all other sills only at ends.

The backs of all ashler, belt courses, cornices, voussoirs, etc., to be sawed or evenly pointed off to the required depths so that proper bond may be had with the brick backing; the same to be done with all sills, jambs, mullions, transoms and lintels.

The several courses of ashler must be of alternate thickness so as to allow of proper bonding with the brick backing.

. All stone work forming jambs to reach back from the face edge of the door and window frames not less than two inches and the window sills and lintels to be two inches deeper. Cut jambs and lintels to clear the window frames and boxes. No vertical joints to show in reveals or jambs.

No patching or hiding of defects will be allowed, and all defective stone work will be rejected. No lewis holes to be exposed or tool or saw mark to show on any rubbed work.

All stone must be perfectly clean when set. As soon as stone is cut it shall be protected from dirt and injury. Should any stone become splashed with dirt during rains or otherwise, it must be washed clean.

All horizontal courses to be kept perfectly true and level, and the walls to be built uniformly.

All projecting courses must have sufficient bearing on the walls to insure stability and must rest on the wall at least equal to one inch more than their projection, or to be built into the wall sufficiently for the weight in the wall to overbalance the weight of projection. The outside bearing point to be at least one half inch from the face.

All stone must be set on their natural quary beds.

Leave under window sills in Office and Court Room stories, openings for fresh air inlets to radiators.

In setting stone the mortar must be raked out one-half inch from the face, the joints kept clear to this depth to receive the pointing on completion of the work.

No stone to be set on another until the stone already set is backed up with brick and made level, unless the top stone is smaller. No third course to be set until the others are backed up and bonded under any circumstances.

Through stone will be built in the walls as shown; the same must be dressed to dimensions and bedded close, the backs to be flush with the inside face of walls.

At projections, washes, copings, etc., there will be saddle joints wherever required, all to be properly calked as may be directed.

The contractor must protect all stone work from injury at all times so as to prevent any possible chance of injury during construction. All projections, angles, mouldings, sills, steps, copings, door and window jambs, etc., must be protected by boards from injury during construction, and the work, as it progresses, must be kept covered with boards when exposed to rain, etc.

Cut all required holes and beds for iron work, raggles for all flashings and other work for roofs, gutters, down spouts, etc., that may be necessary and required.

The contractor is to supply all dowels, cramps, anchors, etc., as required, make the sinkings for same, thoroughly imbed them in mortar, and properly build them in.

It will be required that all stones be properly cramped together and anchored to the brick backing with heavy galvanized wrought iron cramps, anchors, dowels, and pins.

Cramps for coping, balustrades, etc., to be two inches wide by one-half inch thick, of required lengths and ends turned down square one inch, let into stone one-half inch below surface, set in lead, and groove filled flush over with neat Lafarge Portland cement, colored to match the stone.

All work in pediments and tower must be anchored and cramped in a thorough manner.

The size and position of the stone will govern the size and number of the anchors and cramps to each stone as determined and directed by the Architect. None to be less in size than one and one-quarter inch by one-quarter inch, and of the length required by the different thickness of walls. Two or more anchors to each stone where required.

The cornices to be thoroughly tied together and to the brick and iron work.

Anchors for fastening stone to brick backing to be turned down into stone one inch, and to extend into brick work not less than enight inches, and turn up two inches.

Wherever required anchors shall extend to the inside face of walls.

All vertical stone are to be doweled. Stone dowels to be used in columns wherever required and of the necessary size.

The bases, dies and caps of posts, shafts and caps of columns, caps of balustrades, coping, etc., to be doweled and tenoned as will be shown, and all to be anchored properly together.

All cramps and anchors must be dipped while hot into asphalt, and cramps and dowels set in moulten lead.

All stone work to be bedded and pointed in mortar composed of Lafarge Portland cement one part, lime putty one part, and clean sharp sand three parts. Before backing is done the entire back of all stone work to be plastered a thick coat of same mortar, all joints to be flushed full and every precaution taken to prevent discoloration of stone from the cement backing.

The corner stone will be placed in the northeast corner

of the building as shown on Court and Main street elevations. The north and east surface of the stone to have inscriptions with letters sunk "\'" shaped. The contractor to do the lettering following the texts furnished by the Architect. The stone to have dressed margins and faces as may be required. Cut a cavity in the bottom bed to receive a copper box ten or twelve inches square on all sides. The contractor to furnish and provide all requisite power, machinery, etc., attending the ceremonies of laying the corner stone. The corner stone to be the full thickness of the wall. **1619450**

There will be no rubbed work in the Ground Story, except the entablatures over entrances and portions of the granite already specified and what may be required for the corner stone.

The moulded course on top of the granite base will be tooled, the narrow plain surfaces vertically and the double curved member horizontally, all ten cut. The ashler of Ground Story to have one and one-half inch tooled margins and fine bush-hammered centers. The rabbits and sides of projections of each course and keys, also jambs and soffits of the windows to be tooled ten cut work. The moulding forming cornice of Ground Story will be finished in the same manner as the moulding on top of granite base. The entablatures over entrances will be tooled, rubbed and carved as shown and as may be directed.

The sub-plinths under bases of columns, pilasters and corner piers will have one and one-half inch tooled margins, ten-cut both face and sides, and fine bush-hammered centers. The plain surface under moulded window sills will be tooled vertically. In these stone openings for fresh air inlets will be cut as before specified. This course over entrances will be paneled as shown. The moulded sill course will be tooled and the window sill on top will be tooled vertically and the washes of both will be rubbed.

The plinths of bases of columns, pilasters and corner piers and the imposts and small columns over entrances will be tooled vertically.

All ashler between window finish and columns and pilasters will be tooled courses alternately vertical and horizontal, eight-cut. The courses forming corner piers, between their bases and caps, will be tooled margins with fine bush-hammered centers.

All columns, pilasters, friezes, facias, mouldings, jambs and reveals will be fine rubbed.

All other work will be either sawed, fine-rubbed, carved and sculptured as shown, or as may be required and directed.

All moulded work to be well defined, of the dimensions, design and profile as shown and to be cut clear, sharp and distinct.

The contractor to furnish full size models in plaster of all moulded work that may be required, before executing.

All projecting mouldings to be properly weathered, and to have drip cut on under side, and sills to have seats cut for jambs and mullions where shown.

The shafts of columns to have flutings and entasis as shown.

The large arches over entrances will be lowered and the main entablature raised, permitting the architrave and frieze of the entablature to continue over the arches. The sections, details and figures showing this change from the elevations. Inscriptions to be cut on the continued frieze over arches in "V" shaped letters. There will be inscriptions also in the panels under arches and in the larger panels under main pediments as indicated on the elevations. The subject of all inscriptions will be furnished the contractor by the Architect, the contractor doing all the laying out and cutting. This includes the corner stone as well as lettering the small panels under all of the circular windows in Court Room story, between pilasters.

Carving will be done where indicated on elevations and details, except the panels in balustrade which will be sculpture instead of carving as indicated on elevations.

The sculptural work will consist of the portrait busts in circular panels between pilasters on each side of the large arches, and the spandrels of the arches, Court street arch spandrels included, the figures on each side of the panel under arches, the tympanum of all pediments, the four smaller to be sculptured instead of carved as shown,—the figures under the four large pediments, panels between figures, not including inscription panels, and all panels in balustrade over the main entablature, instead of carving as mentioned above. All of this work to represent such personages, characters and events as the Commissioners and Architect shall select and designate and must be done in the most skillful and artistic manner by none but reputable and high class sculptors from plaster models of both full size and approved scale furnished by the contractor and submitted to the Architect for approval before executing. The sculpture must be in full and high relief as its position in the building and the subjects it represents require. The sculpture work may be let separate and distinct from all other portions of the work, but shall remain under the general contractors responsibility in every way provided the work is awarded him. Should the work not be contracted for by the general contractor doing the balance of the work, such general contractor shall furnish, provide and place all the proper material in the walls for the execution of such sculpture work which may be executed at any time after such material is placed in the walls, shall be properly placed and secured and of the proper form and dimensions the same as if the work was to be executed and finished under the contract of the general contract.

All carving as indicated on drawings, except that are specified to be sculpture, is included in the work of the general contractor, including the chimeras at sides of clock dials.

The contractor must first obtain the approval of the Architect before appointing any carvers, as none but thoroughly competent carvers will be permitted to do the work. If, in the opinion of the Architect, those chosen by either the contractor or Architect, or both, are found to be incompetent, they shall be dismissed from the work and acceptable carvers employed by the contractor.

All carving to be cut with spirit and feeling, not mechanically, well accented, clear, distinct and sharp, with good relief and in the most perfect manner and strict accordance with full size plaster models.

The contractor shall have full size plaster models made of all carved work, submit the same to the Architect for his approval before executing.

In all places where possible, the stone for carving to be left rough when set in the wall, and the carving done after the outside construction work is finished.

The stone work for the Power Station is included in this work, the conditions for the foregoing work covering the same. All to be of the same quality of stone, consisting of door steps and sills, water table, window sills, two columns, bases and caps for columns and pilasters, the pilasters being

of brick, the imposts, jambs and reveals of east entrance, and
arch and spandrels over same. The water table to be cut
with a wash, the sills with lugs, all to be rubbed. The caps,
bases, columns, etc., all to be neatly finished.

The stone work for the Smoke Stack is also included as
specified under its own heading.

After the exterior of the buildings are completed in every
particular, all of the exposed stone work must be thoroughly
cleaned from the top to bottom. All exposed joints shall
be thoroughly raked out clean to the depth of one-half inch,
moistened with water, packed solid and full and pointed
with lime putty mortar colored to match the stone. Said
mortar to be composed of fresh burned lime and marble
dust or fine lake sand. Pointing to be finished flush and
concave where directed. The joints of all coping and projecting stone to be packed solid and full, pointed flush with
neat Lefarge Portland cement and made absolutely water
tight. The granite work to be pointed up solid and full
with neat Lefarge Portland cement, colored to match.

All dirt and stains to be removed from the stone, and the
whole left in a perfectly clean and finished condition. All
cleaning and pointing to be done from swinging scaffolds
supported by derricks, no bracket scaffolding will be allowed
and no scaffolds to be hung from gutters or other projecting
stone work in any event.

ARCHITECTURAL TERRA COTTA.

The architectural terra cotta work will consist of the main
dome of the tower, the cornice and lantern above to the underside of the ball. Also terra cotta work for the Power
Station as denoted on drawings for that building, which
consists of all cornices, balustrades, pediment, ridging, etc.
All to be according to details to be given and as shown on
the various drawings.

The tourelle for the spiral stairs at side of tower is also
included in the terra cotta work. All to be as shown.

All moulded work to be made from full size drawings and
the ornamental work to be made from full size models furnished by the contractor and approved by the Architect.

The work to have such ornamentation as indicated, well
modeled and equal in quality and effective modeling to the
best work done by the New York Terra Cotta and Perth

Amboy Co's, to match in color with the stone used in the Court House and the brick in the Power Station.

All the terra cotta to be best quality, burned straight and even, free from kiln cracks or other defects; no spalled, chipped, glazed, or warped pieces will be accepted; to be moulded with sufficient stiffening ribs and be evenly burned to uniform color for the respective buildings.

Wherever practicable, the hollow portions to be filled in solid with brick and cement mortar.

All the terra cotta copings, projecting courses and lintels to be bedded and jointed in cement mortar, and as soon as set, all joints to be raked out to a depth of one-half inch.

All other work, in Power Station, to be set in pulp mortar as specified for facing.

All joints must be regular and not exceed one-eighth of an, inch in thickness.

The terra cotta to be made to lap for all weathering joints.

The terra cotta to be jointed generally as shown on the drawings; but before getting out the work, the contractor must submit his proposed jointing to the Architect for approval.

All the terra cotta for the main dome and dome of lantern must be made with lapping joints as directed and the joints made thoroughly water tight. All to be thoroughly imbedded in cement mortar.

All the terra cotta to be put up in the most approved manner and that for the Court House bedded in mortar made of one part best German Portland cement and three parts clean sharp sand.

All columns not having metal column cores to be carefully constructed of tile.

All work to be properly secured to the backing, wherever any comes behind it, and tied down with the necessary rods, anchors and other supports to the iron frame work. Rods, anchors, and other supports to pass through the backing wherever required and secured to the iron frame work.

All cramps, rods and anchors, to be galvanized wrought iron as specified for stone work.

The terra cotta set in cement mortar as hereinbefore specified, and pointed as specified for stone work.

All pointing mortar to be colored to match the terra cotta, and be especially prepared to render it impervious to moisture.

All terra cotta to be cleaned down with a weak solution of muriatic acid, care being taken that no acid gets on the stone.

Duplicate samples of a section showing the color as it comes from the kiln must be forwarded to the Architect for approval, by express, charges prepaid, before placing any in the buildings.

STRUCTURAL IRON WORK.

For the dimensions, sizes, and the arrangement of the structural iron for this work particular reference shall be had to the drawings. These drawings, with such writings and details as may be upon them shall be considered as part of and illustrations to these specifications.

The factor of safety for steel shall not be less then four, and for cast columns shall not be less than ten.

The contractor to furnish and put in place complete, all steel and iron work shown on the drawings or called for by these specifications.

All workmanship shall be of the best. All abutting surfaces of compression members shall be planed and turned to even bearings so that they will be in contact throughout.

All material shall be of steel unless otherwise noted.

All bolts, tie-rods, T's, anchors and continuous wall plates shall be of wrought iron, shoes for columns and bearings plates shall be of cast iron.

All beams resting on masonry shall have pin and strap anchors bolted to beams.

The bearing plates for girders and beams and all continuous wall plates will be properly set in place and the contractor will be held responsible for their correctness. All bearing plates for girders and beams shall be truncated as detailed on drawings and no plates will be received or allowed to go in the work unless so constituted as shown.

Where two or more beams, or channels or beams and channels are shown to be set together, they shall be provided with cast iron separators, spaced not over five feet centers. One separator shall be set at each bearing and the others spaced at uniform centers as far as practicable. The seperators shall not be less than three-quarters of an inch thick, cast to fit profile of beams or channels exactly, and shall have two bolts to each separator where depth of beam or channel is twelve inches or over.

The entire framing shall be done with standard connections. Where special framing is required, rivets shall not be strained more than nine thousand pounds per square inch single shear. Details of special connection must be approved by the Architect before the work is executed.

The T-irons supported on the roof beams shall be set to the centers indicated and fastened to the beams with malleable iron clamps, driven firmly to place.

The beams for skylight supports and all upright and horizontal angles around skylights shall be furnished and set by the contractor.

The contractor shall furnish and set in place the T-iron supports for all vaulted or groined and suspended ceilings, and unless specially otherwise noted shall furnish all one inch by one inch T's for support of ceiling tile or expanded metal lath. For all vaulted or groined ceilings and for ceiling lights in Court Rooms Nos. One and Four, the supporting T's shall be accurately bent to the various curves required for the work.

All channels and angles around ceiling lights in Law-Library and Court Room No. One and around light well in Dome, and all beams, channels and angles in ceiling light in Dome shall be accurately bent to the various curves as shown on plans and sections.

Cast iron columns shall be cast vertically and the ends planed off at right angles to the axis of the column. The shells shall be of uniform thickness of metal, smooth and sound, without cold shuts, lumps, scales, blisters, sand holes or other imperfections, truly cylindrical, of full diameter and with interior and exterior surfaces concentric. In order that thickness of columns may be callipered, holes shall be drilled by the contractor in the columns at such points as the Architect may direct.

The steel shall be what is known as "medium steel," and may be made by either the Bessemer or Open Hearth process. It must be uniform in quality and must not contain over .10 of one per cent. of phosphorus. The steel shall have an ultimate strength of not less than sixty thousand pounds per square inch. It shall have an elastic limit of not less than one-half the ultimate strength, an elongation of not less than twenty-five per cent. in eight inches and a reduction of area of not less than forty-five per cent at point of fracture.

All blooms, billets, or slabs shall be examined for surface

defects, flaws or blow holes, before rolling into finished sections, and such chippings and alterations made as will insure perfect solidity in the rolled sections.

A test from the finished material will be required, representing each blow or cast, in case the blows or casts, from which the blooms, slabs or billets in any reheating furnace charge are taken, have been tested, a test representing the furnace heat will be required, and must conform to the requirements heretofore enumerated.

A duplicate test piece from each blow or cast and furnace heat will be required, and it must stand cold bending one hundred and eighty degrees over a mandril, the diameter of which is equal to one and one-half times the original thickness of the specimen, without showing sign of rupture on either corner or concave side of curve.

After being heated to a dark cherry and quenched in water one hundered and eighty degress Fah. it must stand bending as before.

The original blow or cast number must be stamped on each ingot from said blow or cast, and this same number together with the furnace heat number, must be stamped on each piece of the finished material from said blow cast at furnace heat.

No steel beam, channel or angle shall be heated in a forge or other fire after being rolled, but shall be worked cold unless subsequently annealed.

In all cases when a steel piece, in which the full strength is required, has been partially heated, the whole piece must be subsequently annealed.

All bends in steel must be made cold, or if the degree of curvature is so great as to require heating, the whole piece must be subsequently annealed.

Steel for rivets throughout this structure shall have an ultimate strength of not less than fifty-six thousand nor more than sixty-two thousand pounds per square inch. An elastic limit of not less than thirty thousand pounds per square inch; an elongation of not less than twenty-five per cent. in eight inches, and a reduction of area of not less than fifty per cent. at point of fracture.

Specimens from the original bar must stand cold bending one hundred and eighty degress, and close down on themselves without sign of fracture on corner side of curve. Specimens must stand cold hammering to one-third their original thickness without flawing or cracking, and must stand quenching and bending as specified for rolled sections.

Where wrought iron is required by plans and specifications, it shall be tough, fibrous, and uniform in quality and shall have an elastic limit of not less than twenty-six thousand pounds per square inch. It shall be thoroughly welded during the rolling, and free from injurious seams, blisters, buckels, cinders or imperfect edges.

When tested in small specimens the iron in no case shall show an ultimate strength of less than fifty thousand pounds per square inch and shall elongate eighteen per cent. in eight inches.

The same sized specimens taken from angles or other shaped iron shall have an ultimate strength of not less than fifty thousand pounds per square inch, and shall elongate not less than fifteen per cent. in eight inches.

All iron and specimens from plate angle and shape iron must bend cold for ninety degrees to a curve whose diameter is not over twice the thickness of piece without showing sign of fracture.

When nicked on one side and bent by a blow from sledge, the fracture must be fibrous, showing but few crystaline specks.

Cast iron shall be the best quality of metal for the purpose. Castings shall be clean and free from defects of every kind, boldly filleted at the angles, and with arrises sharp and perfect. Cast iron must stand the following test: A bar one inch square, five feet long, four feet six inches between bearings, shall support a center load of five hundred and fifty pounds without sign of fracture.

In all work the diameter of the punch shall not exceed by more than one sixteenth the diameter of rivits used. Rivet holes must be accurately spaced; the use of drift pins will not be allowed, except for bringing together the several parts forming a member, and they must not be driven with such force as to destroy the metal about the holes. If the holes must be enlarged to admit the rivet, they must be reamed.

All rivets in this work shall be accurately spaced, so that upon assembling of adjacent pieces a cold rivet of the size intended for the work can be inserted in the holes.

Drifting will not be allowed.

Rivets with crooked heads or heads not centrally located on the shaft, or loose under the heads or in their length, must be cut out. When it is found necessary to cut out rivets in steel work, it shall be done in a way not to injure the material. In case of injury to the material the hole shall be reamed again or material replaced at the option of the Architect, and at the expense of the contractor.

All rivet holes must be so accurately punched that when the several parts forming a member are assembled, a rivet one-sixteenth less than the diameter of the hole can enter hot without reaming, drawing, or straining the material by drifts.

The rivets when driven must completely fill the holes, and the heads must be hemispherical and of uniform size for the same size rivets throughout the work. The heads must be well formed and full, and concentric with the rivet holes.

Wherever posible, all rivets must be machine driven, and by direct acting power machines, which are capable of holding onto the rivets after upsetting is completed.

All iron work must receive a coat of pure linseed oil at the rolling mills just before being loaded on the cars, and all material of every kind and description must be thoroughly cleaned of rust and dirt and receive one coat of pure boiled linseed oil before any work is done on it, and all finished material to be recoated with same oil when former coat has been injured by heating or otherwise.

All finished material must receive one complete coat of No. 30 Superior Graphite paint and all pins, pin holes, planed surfaces and ends of columns must be coated with white lead and tallow before leaving the shop.

All surfaces inaccessible after erection must be painted one coat before erection, and all surface inaccessible after assembling must be painted one coat before assembling, in addition to the coat before leaving the shop.

All surfaces accessible after completion must have one complete coat in addition to the former coat.

All bolts used in erection and remaining permanently in the building must be dipped in Graphite paint before being placed in position.

All painting must be done on dry surfaces and preferably warm ones. All dirt and foreign matter of any kind must be removed from the iron before oiling and painting.

All paint used must be No. 30 Superior Graphite paint

prepared and mixed by the Detroit Graphite Manufacturing Co., of Detroit, Mich.

All finished material must be given a thorough surface inspection, and must be perfectly straight, clean, smooth and free from flaws, cracks, cinder pockets and other imperfections.

The contractor must furnish without charge full and ample means for the inspection and testing of all rolled, forged, or cast material for this work. He shall furnish without charge such prepared specimens of the several kinds of iron to be used as may be required to determine their character, and shall admit the Architect or his authorized inspectors to any portion of the mill or shop where work is being done under this contract.

Full size members may be tested at the option of the Architect, but if tested to destruction and proven satisfactory, such material shall be paid for at cost, less its scrap value. If it does not stand the specified tests, it will be rejected material, and be solely at the cost of the contractor.

The contractor shall furnish free of charge, the use of a testing machine capable of testing the specimens. He shall also furnish all necessary assistance in handling and operating the same.

The inspection shall be made as soon as possible after the material is cut to lengths and ready for shipment, and all material must be shipped from rolling mills as soon as possible after inspection. Should any delay occur in shipment of material after inspection, it shall be so piled up in a manner and place as not to be injured by rust or otherwise. Before being piled up it must receive one coat of boiled linseed oil.

The acceptance of material by inspectors will not be considered final, but the right is reserved to reject at any time before the completion and acceptance of the work. Any material which may prove to be defective, and all damage to other work by the removal of faulty iron or steel, together with cost of removing and replacing new iron work must be made good by the contractor.

The iron work must be carefully erected by skilled labor, all connections must be riveted, and the erection must be carried on in such a manner as to avoid any damage to the masonry work, and at such times as will permit the steel work to be most thoroughly incorporated with the balance of the work.

These specifications are intended to call for the best workmanship and design throughout, and the details must be made in conformity with this. All connections will be proportioned for the full strength of the individual pieces without regard to the actual strains occurring in these pieces.

The contractor will be furnished with such additional detail drawings from time to time, as the work progresses, as may be necessary to clearly show the character of the connections. The contractor will then prepare all shop drawings for the work and make all measurements at the building necessary to verify his work and the general drawings.

All drawings prepared by the contractor must be duly approved in writing by the Architect before commencing to get out the work at the shop.

Such approval will be only as to general detail, and the contractor shall be responsible for all fits and clearances.

Any work showing obvious defects in strength or arrangement from that called for by approved drawings may be condemned at any time prior to the final acceptance of the work as a whole.

The contractor shall prove up all constructional iron work and notify the Architect of any defects.

There will be extra columns, etc., as noted on Foundation plan.

When partitions or columns do not come directly over floor beams a sill composed of an I beam or suitable plates shall in every case be used to properly distribute the partition load. When partitions need lateral stiffness this is to be given by the use of upright beams or T's. Such will be used in the partitions between corridor doors on Court Room floor, and elsewhere wherever required. This construction is not shown on detail for same but attention is called to the fact that they will be required.

These will consist of four inch I beams built in the tile partitions to stiffen the same, to be placed at each jamb of every opening, to have top and bottom plates long enough to be bolted or riveted to the floor and ceiling beams, the beams to be secured to the plates by angles on each side, top and bottom.

Provide for all vaulted ceilings under stairs, etc., as shown.

The ceilings in passage ways from Commissioners' Court Room to be lower than others, as shown.

All openings over vaulted ceilings, hollow or double walls, difference in levels of ceilings as above stated, and over large pendatives in rotunda, to be provided so that they may be floored over and protected.

There will be seven panels of ceiling lights over each communicating corridor instead of five as shown.

It is desired and intended that the deck roofs on each side of the central roofs of the building be raised about two feet in order to gain more height for the Jury rooms in attic. This the contractor will bear in mind and make all the necessary provisions for doing without any additional expense to the county.

Furnish and construct flag staffs as shown, to be made of wrought iron pipe of different diameters to form diminution, to be securely and neatly constructed at joints and braced in a rigid manner from the attic in cast iron shoes and properly flashed with 18-oz copper at roof lines. The tops to have copper balls and eagles gilded with xxxx gold leaf, mounted on alluminum bronze rods with alluminum bronze collars and each fitted with two lignumvitae sheaves and coper line cleats placed where directed. To have the necessary length of halyards.

Furnish and secure in place awning fasteners for all east, south and west windows.

Furnish and properly secure in place guard rails around all ceiling lights in attic, made of gas pipe, three feet high, with intermediate balusters twelve inches from centers.

There will also be a railing or balustrade around the ceiling light of Rotunda as shown, with uprights or balusters spaced as above mentioned with an intermediate rail as shown. Around this balustrade will be neatly and properly secured wire netting of approved guage and mesh. Wire netting will also extend up all stairs from this floor to and including the spiral stairs to the Lantern of Dome.

The large window frames and sash in Tower will be constructed for glazing and made in a neat and acceptable manner.

There will be a ladder from the floor of Lantern to the dome of same for the purpose of climbing to the electric

light fixtures for illuminating the statue. The ladder to be
made of gas pipe rails with solid iron roads and securely
fastened at both ends. There will also be ladders on two
sides of the tower running down the hollow walls of tower
over arches from the balcony floor under inner dome to the
attic over ground stair case hall and the Law Library.
Provide open work iron doors in attic at corner of tower
and in partition as shown.

The framing for the glass in eye of light dome over ceiling
of Rotunda and those over the Court Rooms will not be included in this work as it is intended to include that portion
of the work with the glazing. All work for that purpose that
is detailed with the other iron work will, however, be included in this work.

The retaining cell shown on Ground Floor will be constructed of wrought iron bars and rods carbonized, and lined
with boiler iron. The boiler iron door, walls and ceiling
will be three-sixteenths of an inch thick, to this will be rivited bar iron two and one-half inches by one-half inch, six
inches between centers, rivets one-half inch, countersunk on both sides. The angles will be formed by three
inch angel iron riveted countersunk. The inside door will
be made of bars and rods, the bars of the size above mentioned and the rods one and one-quarter inch diameter, to be
countersunk at ends, the rods to be bent at ends forming
corners of door. The horizontal bars will be bent on ends
riveted to side of door, rivets countersunk, placed eight
inches from centers, the perpendicular rods six inches from
centers. The door to be fastened to the cell with strong
hinges and have Yale lock. The outside door will be plate
or boiler iron, paneled with ornamental cast iron moulding
around panels. To have hinges nickle plated and Yale
lock. All of this work to be made in a substantial manner
and of neat appearance. In carbonizing the bars and rods
there should be to sixteen hundred pounds of iron, four
pounds of Prussiate of Potash, two bushels of burned bone,
four bushels of charcoal, one-half bushel of rock salt, put
iron in perfectly air tight cast iron box. Heat from six to
ten hours. Immerse in cold water.

There will be two stairs to Basement, one from the Engineer's room and one from the Janitor's room. To be without risors. Two stringers of strength equal to a seven inch
bulb angle eighteen and one-quarter pounds per foot, treads

one-half inch metal by eight inches wide. Diamond cut.
To have hand rails, etc., complete.

The stairs leading from the Public Corridor on Court
Room floor to Attic, the short stairs in attic at each end of
balcony and the spiral stairs leading from Attic to Tower to
be put together in the best manner with brackets, lugs,
flanges, etc., and to be securely riveted, screwed and bolted
together. All to have diamond cut treds and plain risors;
rail to be of gas pipe, balustrades of wrought iron.

All of these stairs to have cast iron wall bases.

Provide for the additional heights of floors in attic as noted
on attic plan.

All stairs to be strongly supported on steel strings and
carriages. The ends of strings to be bent to the proper
curve. To have lugs for supporting the treads.

Provide and set all the "I" beams, channels, "T's" or angles
that may be required to support the landings, platforms,
strings, etc., of the different stairs.

All stairs in tower leading to lantern to be carefully constructed. The platforms, landings and floors of balconies
to be of cast iron, the balustrades of wrought iron, and gas
pipe rails. All to have wire guards as before specified. The
stairs to have plain cast iron risors and all treads and floors
to be diamond cut. This work to be securely bracketed and
bolted to the structural work. The balcony above mentioned runs around to all clock dials.

All stairs on Ground, Office and Court Room Floors
and those leading to the Jury Rooms in Attic are to have
marble treads and risors. The iron work or these stairs
will be constructed accordingly.

The soffits of the main stairs leading from Ground Floor
to Office Floor, and from Office Floor to Court Room Floor,
and from Court Rooms to Jury Rooms in Attic, and the
private stairs leading from the Sheriff's Office to Court
Room Floor will be furred with iron furring set twelve inches
on centers, and covered with stiff metal lath for plaster. The
furring is included in this work, and must be securely
fastened to the carriages of the stairs.

The ceilings of Jury Rooms and closets in attic will be suspended from the iron roofs by angle irons and will be made
of angle and "T" bars and furred with iron furring set
twelve inches on centers. All the above furring and lathing

to be done in consultation with the Architect, and according to his direction.

All the stiff metal lath to be good, heavy lath, approved by the Architect.

All furring that is not fully explained on the drawings, to be carefully and substantially done, so as to leave all work secure and amply strong.

Provide cast iron pockets and covers for electric light fixtures under balcony floor near ceiling of rotunda, covers to be hinged and clasped. The approximate size can be obtained from the detail drawing of rotunda and the number from the Court Room Story ceiling plan.

Channeled steel thresholds of approved profiles to be provided for all exterior doors of Court House and Power Station, fastened with countersunk head expansion bolts, the thresholds to be drilled to receive the bolts of standing leaves of doors.

Steel beams to be provided for coal areas for Power Station, as shown, also covers for coal chutes, frames for same to be framed between the beams in a secure manner. Covers to have staples, chains and bar fastenings on inside.

Also provide all necessary rods, bolts, stirrups, etc., for the wooden roof construction of the Power Station.

Provide all iron work for Tunnel and Smoke Stack as specified under their respective headings, and shown on drawings for same.

Provide scuttle, curb and lid for floor of lantern of dome, the lid to be hung on two heavy wrought iron hinges, and have clasp, staple and brass padlock with two keys, and provided with approved quadrant opener properly secured. All to be of iron of the lightest possible construction commensurate with safety. Provide frames, hinges, locks, etc., complete for four doors out to the four deck roofs from side walls under the main roofs, to be covered with copper as hereinafter specified. The entire floor of lantern to be of iron. Provide necessary means of access to roofs.

Particular attention is called to the construction of the iron work for marble stairs. The contractor should familiarize himself with that portion of the work and execute the iron work to conform to the marble work.

ORNAMENTAL IRON WORK.

The private stairs leading from Sheriff's Office to Court

Room Floor, the stairs in Auditor's Office to Ground Floor, the stairs from Court Rooms No. Two and No. Three to the Jury Rooms in attic, and the two spiral stairs in Law Library to be of design similar to that shown on detail No. 19. All rails, newel posts and facias at side of stairs and around the floors will be of cast iron. The facias will be the full thickness of the floors and turn under the soffits and celings with mouldings to form a finish on the under side. The newels and facias will be ornamented as shown. The base above the steps will be of iron.

Provide for additional height of floors in attic as noted on plans.

All of these stairs will have marble treads and risors except the spiral stairs which will have marble treads only, the risors to be of cast iron open work of ornamental design on both sides. The balustrades of all these stairs to be of wrought iron, of ornamental design. The nosing and floor pieces outside the balustrades will be cast iron. The facias will extend around the well at ceiling of stairways to attic Jury Rooms.

The panels in marble balustrades around well holes in the Office and Court Room Floors of Rotunda to be of wrought iron of approved design similar to that denoted on details.

The balustrade around balcony in Law Library and facia around the edge of balcony floor to correspond with the stair work above specified, including the nosing and floor pieces.

The facias around well holes in the above mentioned rotunda, to be of cast iron and as shown on details.

The public and private elevator fronts and the side of private elevator to be of wrought iron of similar design as shown on detail drawing for public elevators. The public elevator fronts to have wrought iron grills over each door.

The balcony over the grand stair case and the balcony around rotunda under inner dome to be of wrought iron, the floor pieces, nosings and facias of same, the cornice under inner dome, the ribs, panels, etc., forming frame work for glass panels of inner dome, and the facia and mouldings of the eye of said dome to be cast iron of ornamental design.

There will be cast iron mouldings of ornamental design around the ceiling lights of Court Rooms, Law Library, Jury Rooms in Attic, over private stairs, over the ends of judiciary corridor and all other ceiling lights shown on the drawings. To be open work wherever required for ventilation. The balcony of lantern on tower to be wrought iron of ornamental design. The balcony shown on elevations being plain to prevent confusion with other work.

The crown member of the cornice of dome of tower will be cast iron of ornamental design, to have lions' heads and as otherwise shown on detail drawings.

The floor and stair balustrade in Power Station to be of wrought iron of ornamental design as shown, to have floor pieces, nosings and facias of cast iron. Risors paneled, treads diamond cut.

The Calhoun, Main, Court and Berry street entrance doors will be of designs as shown and noted on detail drawings, together with frames, transoms, sash, etc., as shown. All to be of metal, constructed on wrought iron frames and securely bolted together in the very best manner.

All exposed finished metal work specified above to be heavy copper electro-plated or deposited on the iron, to be then refinished as if it were solid bronze, to be from the shops of Poulsen & Eger, John Williams, Winslow Bros., or the Jackson Architectural Iron Works. The models to be well made, and the castings carefully done and finished so as to be perfectly smooth and free from all roughness.

CONSTRUCTIONAL TERRA COTTA.

All floor arches, ceilings, roofs, partitions, beam fire-proofing, and all fire-proofing of iron to be the best quality porous terra cotta or hard tile thoroughly and evenly burned, good shapes and surfaces, free of flaws, fractures and all other defects, and from some reliable and well known manufacturer thoroughly satisfactory to the Commissioners and Architect.

Each piece of terra cotta, before burning, to be roughly scored on beds, joints and faces intended to be plastered, to afford suitable key for mortar.

All floor arches to be terra cotta flat arches, either "end construction" or voussoirs, of depths shown and figured on drawings. The soffit line of the arches to be below the bottom flanges of the floor beams and channels, and the skewbacks so constructed and arranged as to fit solid on the flanges of the beams and channels, that the soffits of same will be completely covered by slabs of tile keyed between the skewbacks, or specially made skewbacks meeting at the center of flanges will be acceptable.

No joints to exceed three-sixteenths of an inch in thickness, and cutting and fitting will be allowed only where absolutely necessary, and approved by the Architect. In case the regular forms of arches will not fit, special patterns must be used.

The arches must have as great a strength as the beams will admit of, and so constructed as to derive the greatest amount of strength out of the lightest possible construction.

No brick will be used in keying, but flat thin slabs of tile or slate may be used only where absolutely necessary.

All ceilings to be constructed in the same manner as the floors and with the same kind of materials.

All arches to be set on plank centering, to have sufficient camber to ensure flat and level arches when completed.

The raised floors in closets and elsewhere to be constructed of hollow terra cotta blocks, bedded and jointed in cement mortar on the top of the floor arches. These blocks to be properly cut and fitted over the beam flanges, plumbing and ventilating pipes, and leveled up with cement mortar for proper bedding of the floor tiling and marble slabs under the fixtures.

Special attention is called to the raising of floors in Jury Rooms and closets in attic, and also the floor of stair landings to Jury Rooms, and the stairs leading to tower. These floors are to be raised particularly to clear the ventilating ducts under them, and this point is also noted on the attic floor plans.

These floors are to be raised in the following manner: Build four inch tile walls to the required heights, sixteen inches from centers, over the portion to be raised, and lay on top of these walls three inch book tile, four and one-half inches below the finished floor level. These floors to be built in connection with the ventilating pipes which they are to inclose.

The floors in the recesses of alcoves in gallery of Law Library will be raised in the same manner, in order to keep the floors on the same level throughout these alcoves.

Floor over all hollow spaces in attic and elsewhere caused by double or hollow walls, groined or vaulted ceilings and pendentives, and the lowering of ceilings over passage ways from Commissioners' Court Room.

The hollow or double walls in Treasurer's vault, the hollow or double walls or partitions between Court Rooms Nos. One and Four, and the communicating corridors adjoining them, and all similar hollow or double walls or partitions throughout the building, will be anchored together with pieces of heavy wire cloth, four inches wide, and of required length, one anchor for every three square feet of wall.

All partitions throughout the building will be constructed of hollow terra cotta blocks set on the floor beams or supplementary I beams, well secured to walls, floor and ceiling. To be of thicknesses shown on drawings. Portions of the partitions to be stiffened by I beams. Partitions to be secured to abutting brick walls, with large spikes driven into the joints of the brick work, and all partitions wedged tightly to ceilings and spiked. Porous terra cotta blocks to be built in jambs for securing frames. Arches to be formed of blocks over all openings. The contractor to furnish and set permanent frames to guide the work.

All partitions of Jury Rooms and closets in attic to be built up under the roof, and not as shown on the drawings.

Particular attention is called to the furring required throughout the building, and shown on plans, sections and details. All must be carefully and accurately laid out and constructed in the most rigid manner, of the very best material and of the dimensions required. Do all furring wherever necessary to conform to plan of rooms and alcoves, or to form piers, columns, pilasters and panels; also all furring in closets and lavatories for ventilating pipes and flues. All exterior walls to be furred on the inside of rooms and closets with two inch porous terra cotta furring, secured to walls with cement mortar and also with large spikes with metal caps, driven through the furring into joints of the brick work.

All the cast iron columns throughout the building to be jacketed from bottom to top in the best manner with approved terra cotta column covering, not less than two inches thick, with ribs on the inside, allowing one-half inch space between iron and terra cotta.

All exposed portions of the ground floor iron work in basement ceiling to be entirely covered with terra cotta fireproofing, not less than seven-eights of an inch thick. The several sections of furring to be properly secured in place in an approved manner.

The entire roof and domes of tower and lantern to be covered with terra cotta as shown on drawings. Construct saddles to form pitches to down spouts as shown. All of this work to be secured to the iron work in the most secure manner, and be of an even surface throughout, and all joints thoroughly filled with cement mortar.

All columns, beams, girders, etc., not built into walls, to be thoroughly protected by terra cotta covering, secured to said iron work, completely covering it at every point and thoroughly secured, to be of the best quality, and set in the best and most approved manner.

All constructional and fire proofing terra cotta to be set in cement mortar composed of one part Portland cement, two parts clean sharp sand. Same cement and tests as specified for foundation. All to be thoroughly bedded and the joints completely filled.

The contractor shall do all cutting required for fitting his work, and do all cutting and fitting that may be required by plumbers and other contractors, including his own work, and make all necessary repairs without extra charge to the county. He shall furnish all centerings, materials, water and appliances of every kind required for the execution of the work.

It will be required that all of the terra cotta shall consist of nothing but the best and latest improved material and construction, and executed by experienced mechanics, in the best manner possible.

The roof must be entirely completed before commencing any other portions of this work.

TILE ROOFING.

The four main roofs of the Court House will be covered with Akron natural red terra cotta roofing tile, eight inches by sixteen inches in size, and uniformly one-quarter of an inch thick, laid with a lap of three inches of the third over the first, showing six and one-half inches to the weather. All to be well burned and uniform in size, color and texture; the surfaces smooth and out of wind, edges and tails straight and square, and the corners full.

Courses at eaves, ridges and skylights to be doubled

and cemented, and all courses laid perfectly straight and horizontal, and the bond perfect.

Each and every tile to be securely fastened through countersunk holes at waists with copper bolts of approved size, and long enough to extend through the concrete covering and the three inch book tile between steel T's. Bolts to have heads to fit countersunk, with nut and washers on under side. All to be bolted tight, and the nuts properly secured from becoming loose. Wherever the iron work of roof may interfere with the use of bolts, strong copper wire must be used and the tile securely tied from the underside of roof. All roofing tile to be thoroughly bedded in Portland cement mortar in the most careful manner, form a true and even surface throughout, and bolt or tie every tile as soon as it is placed.

The surface of concrete on book tile to be well prepared for this work.

The ridging, of corresponding color to the roofing tile, to be made from specially designed pattern, and securely tied to the iron work of roof, and thoroughly bedded in cement mortar and pointed in same, colored to match the ridging.

The sizes of the skylights on these roofs should be taken from the deatils and not the elevations.

SLATE ROOFING.

The four deck roofs of the Court House will be covered with slate tile, to be approved by the Commissioners and Architect, one-half inch thick and eight inches square, thoroughly bedded and completely jointed in actinolite cement. The surface of the concrete on book tile to be well prepared and covered with best five-ply composition, omitting the gravel top coat. The slate to be bedded to form smooth surface throughout, and all saddles, hips, ridges, valleys and runs to outlets and down spouts to be properly formed. Fit close to all walls, skylights, vents and exhaust heads. All joints to be thoroughly pointed and the roof to be made absolutely water tight. This work to extend up the walls sufficiently high for flashing.

The slate must be the best, of uniform color and thickness, the surfaces smooth and out of wind, edges cut straight and square and the corners full.

The roofs of Power Station to be covered with best quality Peach Bottom slate, of uniform color, and split to uniform thickness of one-quarter inch, with the surface smooth and out of wind, edges and tails cut straight and square. The slate to be eight inches by sixteen inches, laid with three inch lap of third over first, showing six and one-half inches to the weather. Each slate to be fastened at waist with two flat head galvanized iron nails. Holes in slate for nails to be drilled and countersunk. The courses at eaves and ridges to be doubled, courses at eaves, valleys and ridges to be bedded in slaters' cement. The lines of the tails of the slate when laid to be straight and horizontal along the eaves, and at valleys cut parallel with valley line. The slate at the hips, valleys, etc., to be cut so that their bond will be uniform with the rest. Cover the roofs under the slate with one layer of best heavy rosin sized sheathing paper, lapped three inches and tacked in place.

COPPER WORK.

The four main roofs will have copper gutters, made of 20-ounce copper. The gutters to be formed over continous T irons as directed, and the upright copper work to be braced and tied with copper bands riveted and bolted or tied at the ends. The entire work forming these gutters to be constructed on wrought iron lookouts, eighteen inches from centers, securely bolted to the iron work.

The gutter openings to be spacious and to extend below the underside, and properly connected with five inch copper down spouts, reaching near the deck roofs. The gutters to be saddled to form runs to these openings as shown. The gutters to run up under the slate eight inches and bolted, or tied, in same manner as slate. Flashings will extend up the roof at tower and pediment walls, and be at least sixteen inches high, and under the slate eight inches, all to be double locked with cap-flashings of copper secured to the raggles and joints in stone work. Flash flag staffs as specified in iron work. All flashings to be 18-ounce copper.

Flashing will extend up and around all the walls above the four deck roofs. To be counter flashed at outer walls in the same manner as called for in tower and pediment flashing. The flashing on the walls under gutters of the main roofs to be double locked to the lower flashings of said gutters. All to be securely fastened to the walls in the most thorough and practical manner.

All the cap-flashing to be built in with the walls at least four inches at the time the walls are laid.

Flash around all skylights, fresh air intake, ventilators and exhaust heads in approved manner.

Cover wrought iron roof doors with copper, and flash openings for same. Doors specified in iron work.

All flashings to extend down over the slate at walls.

Outlets of copper to be properly formed under the slate and extend down to attic, and made absolutely water tight at connections with roof and the wrought iron down pipes.

All outlets on deck roofs to be provided with heavy copper wire domed strainers.

The dimensions of all skylights should be taken from the detail drawings and not the elevations.

Copper screens of twelve mesh and heavy wire must be securely fastened back of each register face of fresh air inlets under the outside windows of the Office and Court Room stories.

All gutters, valleys and flashings for the roofs of the Power Station will be of 16-ounce copper. Connections with the terra cotta cornices to be made in the most approved manner. All valleys and flashings to run to safe distances under the slate, and at walls to be cap-flashed with copper, the caps to be built in with the brick work. Provide copper screens for down spouts.

All the copper to be of the best grade and the work done in the most practical manner.

Copper work for the smoke stack is specified under its own head, and for the flag staffs under iron work.

The four deck roofs will be raised eighteen inches, therefore requiring less copper flashing than shown on the drawings.

SKYLIGHTS.

All skylights shown on the drawings to be either Vaile & Young's, Hayes' Patent or Paradigm, as may be approved by the Commissioners and Architect, to be strongly constructed with the necessary steel framing and ribs, and covered and finished in the best manner. The entire exposed metal parts to be of 18-ounce cold-rolled copper, and glazed with thick best quality ribbed glass.

The skylight over private staircase to have ventilator as specified in Ventilation.

All skylights to be provided with approved condensation gutters, and have netting under them of heavy galvanized iron wire to protect ceiling-lights directly underneath.

Do all necessary framing around skylights that may not be specially shown on drawings.

All skylights must be made to present a neat and finished appearance.

REVOLVING STATUE.

The statue on lantern will be approximately of the size shown, to be proportionate to its height from the ground and the position it is to occupy, and made from full size models, specially designed, to the entire satisfaction of the Architect.

The statue to be constructed on upper and lower ball bearings, the upper bearings to work on the inside of the ball, immediately below the feet of the statue, and the lower bearings, on which the statue is to be placed, about nine feet below in the dome of lantern. The lower bearings to take the entire weight of the statue, and upper bearings to take the side or wind thrusts and hold the statue perpendicular.

The mechanism of this statue must be so adjusted and the statue poised and balanced, that one-half pound wind pressure will cause the statue to revolve.

The statue to be made of 32-ounce copper, joints countersunk and riveted, soldered on both sides, and made perfectly water tight. To be perfectly supported and braced with copper and brass bars on the inside.

The statue to be oxidized to represent antique bronze, and compare in every detail with the finest and best cast bronze work, and must in every way be equal to the finest and best work that can be purchased for the sum of one thousand dollars, permantly set in place on the building to the satisfaction of the Architect.

It is preferred by the Architect that this work shall come from the shop of W. H. Mullins, Salem, Ohio, and the same must be equal in every respect to the very best work there made.

METAL CEILINGS.

The metal ceilings and cornices of entrance vestibule, offices, machinery hall and boiler room of the Power Station, to be of stamped sheet steel, No. 26 B. W. G., contructed in the very best manner on wrought iron frame work and brackets. The ceilings and cornices to be suspended by wrought iron work from the wooden roof construction above. All to be of profiles and paneled as indicated on drawings, and in accordance with details to be supplied. The work to be painted on both sides and before securing in place.

Four-inch down pipes from gutters of the Power Station will be located as shown. To be No. 18 galvanized iron, corrugated and square. To have neat heads, properly constructed under cornices with gutters, and connected at water table with cast iron pipe reaching the drain tile, and connected thereto. The down pipes to be well riveted and soldered, and secured to the walls with approved hold fasts. The cast iron pipe to be secured to the stone water table in the proper manner. The down pipes to be painted before secured in place, and the cast iron pipes dipped in asphaltum. All to be included in this work.

CEMENT FLOORS.

The entire floor surfaces of the Court House basement and the Power Station boiler room and fuel bins will be cemented.

Clean out all rubbish, level off or fill in with soil wherever necessary to the required levels, thoroughly rammed solid and compact at short intervals, between depths of fillings, until the required surface is reached.

Lay foundation of broken stone four inches thick, level and compact. Over this lay concrete bed four inches thick, laid in the same manner and using the same kinds of material as specified for concrete footings.

The finishing coat of these floors to be two inches thick, and composed of two parts cement and three parts clean, crushed granite or furnace slag, free from dust or dirt, properly mixed and spread upon the concrete bed; the finish coat to be dusted with cement, trowled to a hard finish, and the surface slightly indented for foothold. The finish coat to be turned up six inches against walls, piers, columns and foundations for machinery, forming one-half inch thick cement skirting, with splayed top.

The floor around inner dome of tower will have concrete

bed on filling, finishing coat and skirting as specified above, made absolutely water tight.

Make all provisions for contraction and proper drainage, drains to be trapped.

These floors must not be laid until all other work coming in contact with them has been completed.

CONCRETE FILLING.

The top surface of all tile work forming floor arches, ceiling arches, ceiling and roof covering, and all other work herein specified or shown on the drawings, will be covered and leveled up with concrete composed of one part cement, two parts clean sharp sand or coal cinders, to be mixed and used in same manner as specified for concrete footings.

On all floors where mosaic tile are to be laid the top of this concrete shall not come nearer than four and one-half inches to the top of the finished floor surface.

In cases that may occur where the height of the floor arches come to the required level of concrete, or higher, they shall be covered with a coat of cement mortar as directed.

None of the exterior work to be done during freezing weather, nor the interior, unless the building is enclosed and kept heated during the work and after.

All of this work must be thoroughly protected from injury in any way, and any work injured must be taken up and replaced to the satisfaction of the Architect.

The attic surface of the ceilings over the attic Jury Rooms and closets for same will be concreted as directed.

MOSAIC TILE FLOORS.

All floors throughout the entire Court House and Power Station will be mosaic tile, except what is already specified to be of cement. There will be no wood floors in either.

The tile to be those manufactured by the Mosaic Tile Co., Zanesville, Ohio, or its equal, to be approved by the Commissioners.

The tile to be colored clay, burned to a hardness so that they cannot be scratched by hardened steel, and of a density that the water absorbtion does not exceed three per cent., but to be over one per cent., so that the tough clay-like nature has not been superseded by a brittle glass-like structure.

The surface of the tile to be crenelated or indented to afford a foot hold.

The thickness of the tile to be three-quarters of an inch.

The designs to be inlaid in the surface of the tile in colored clays, penetrating one-eighth of an inch.

The designs to be in conformity with the drawings incorporated in the general plans, and noted thereon. These, however, are subject to change made by the Architect.

The tile to be six by six inches square generally, except such narrower strips or fillets as may be necessary to carry the borders or margins to the required widths.

The surface of the back of the tile shall be broken by bosses and indentations to afford the tile a better hold in the mortar.

Before laying the tile the contractor must submit a diagram showing his proposed method of laying and jointing, and after receiving the approval of the Architect, the work to be executed in accordance therewith.

None of this work will be permitted to be commenced until the plastering and other work of like nature is finished.

Before the concrete bed for the tile work is laid, the concrete filling over arches must be thoroughly cleaned off, and all imperfections, if any, remedied. When in the proper condition, and immediately before laying the concrete bed, the filling shall be thoroughly moistened to prevent the absorbtion of the water contained in the concrete bed.

The concrete bed on which the tile are to be laid shall consist of five parts of clean sharp gravel or coal cinders, two parts of clean sharp sand, and one part of best German Portland cement.

The materials shall be mixed thoroughly while dry, and after this, mixed carefully with water. The per centage of water not to exceed ten per cent.

The materials so prepared shall be instantly spread upon the concrete filling and rammed down carefully with small iron rammers, until the concrete bed is closely packed.

Under no consideration shall the mixed material be allowed to remain unused for over one-half hour.

After the concrete is laid it shall be protected against disturbance of its surface, and in places where it is necessary to walk over boards shall be laid.

This concrete bed must be three inches thick, and the

surface one and one-half inch below the finished floor surface, and in all cases shall correspond with it.

All work pertaining to the laying of the concrete bed must be in connection with the tile laying, and immediately before they are laid.

The tile to be laid in mortar which shall consist of one and one-half parts of clean sharp sand and one part of best German Portland cement, mixed with water. The mortar shall on being used be fresh and stiff.

The tile shall be laid in a true and level plain on full beds, and in joints not to exceed one-sixteenth of an inch in thickness, and grouted completely full with clear Portland cement mixed with water.

After the tile are laid the floor must not be used for five days.

All cracks, settlements or damage which may appear in any of the tile floors within one year after the completion thereof from any fault of material or workmanship, shall be made good by the contractor without expense to the county, for which time the contractor shall guarantee all tile floors.

PLASTERING.

All furring required in the Court House and Power Station, that cannot be done with terra cotta, and not specified in the iron work, to be done with proper sized angle or T irons and expanded metal lathing, arranged and made secure for plastering. In the case of domes, pendentives, groins, panels, beams, cornices, etc., to conform as near as possible to their contour, so as not to cause too thick a bed of mortar.

Girder cornices to be furred with wrought iron brackets fastened to the upper flanges of girders, the imitation girders to be furred down with wrought iron brackets fastened to the upper flanges of beams and soffits of floor arches; the brackets to be spaced twelve inches on centers.

All pilasters, columns, capitals, bases, arches, soffits, pediments, and other projections will be furred with iron furring, where terra cotta is impracticable, securely built out, braced, and fastened to the walls.

All furring that is not fully explained on the drawings, to be carefully and substantially done, so as to make all work secure and amply strong.

All chases in walls, openings and partitions to be furred where brick or terra cotta cannot be used.

All furring and furred surfaces throughout the buildings that are to be plastered, to be lathed with expanded metal lathing of No. 24 B. W. G. metal or other metal lathing equal thereto, and approved by the Commissioners and Architect.

The plastering throughout the Court House above the basement, and the two office rooms and water closets in the Power Station, to be best three coat work, hard finish, except where otherwise specified and shown on the drawings to the contrary.

This includes the Jury Rooms and closets in attic, the room in attic at stair landing and the walls and ceilings of stair way leading to balcony around the underside of inner dome, and the balcony above inner dome.

The partitions of the Jury Rooms and closets on the attic side are to be plastered, also the attic side of partitions of Law Library alcoves.

The partitions between Court Rooms over communicating corridors are also to be plastered for the purpose of assisting the lighting of these corridors from the skylights on roof.

The two office rooms in Power Station will be plastered instead of lined with enameled brick as denoted on the drawings, the ceilings to be of metal as hereinbefore specified.

All walls to be swept clean and the brick work and terra cotta work wetted immediately before being plastered.

The plastering to be in two separate coats of brown mortar, finishing coat to be of lime putty and clean, sharp white sand, trowled to a hard smooth finish; no plaster-of-Paris to be used except for the cornices, etc., hereinafter specified.

The mortar for plastering to be composed of best quality, wood burned stone lime, slacked at least fourteen days before using, run through a fine seive, mixed with clean, sharp sand, one-third lime and two-thirds sand.

The first coat on lath work to have one and one-quarter bushels, and the first and second coat on brick work and terra cotta and the second coat on lathed work to have one-half bushel of best quality long ox hair to each barrel of lime, beaten and wetted down before mixing the mortar and after the lime shall have been thoroughly slaked.

The first coat to be put on strong. brought to a fair surface and scratched. After the first coat is dry, plaster screeds five feet apart to be formed on walls and ceilings, properly leveled and plumbed to the required thickness, and the second coat of plaster laid on and filled flush with the screeds, smoothed off with straight edges and brushed.

When the second coat is thoroughly dry the finish coat to be applied.

All angles to be true. ceilings perfectly level, and walls straight and plumb, and the plastering when completed to be clean and free from blisters. cracks and all other defects.

All arches and round corners to be worked to a true radius.

All external angles where left in plaster finish to be protected by one-inch by one-quarter inch iron strap to the height of eight feet above floor, secured at five places in their height, to be securely and entirely imbedded in the mortar.

Plaster to extend behind all wood finish or frames and down to the floor, unless directed or shown on the plans to the contrary. No plastering back of marble lining.

No slacking of lime will be permitted in the buildings.

ORNAMENTAL PLASTER WORK.

All cornices, and capitals except as hereinafter specified. paneled soffits of beams, arches, and ceilings, beam brackets, all ornamental work, panels. mouldings, and enrichments of every kind, to be run or cast in plaster-of-Paris and best fibre, and secured in place in the most thorough manner.

All to be of the profiles and designs shown. to be well and carefully made in strict conformity with full size details or models as the case may be.

All inscriptions to·be properly lettered and the sculptural work to be modeled in the most artistic manner, to represent such personages, characters and events as the Commissioners and Architect shall select and designate.

Do all casting from full size clay models, the models to be made exactly in the position the work is to occupy.

Models of all sculpture, enrichments. ornaments, contours of cornices, and the like, to be submitted to the Architect for his approval before proceeding with the execution of the work.

All models to be specially designed for this work.

All models and designs must be entirely to the satisfaction of the Architect, and any modelers proving to be incompetent in the opinion of the Architect, shall be replaced with competent modelers and employed by the contractor.

Variations of different lengths and other dimensions of the work must be provided for and models must be made to suit the requirements. All mitres, continuations and returns must be done in the most approved manner, as no patching of any kind will be permitted.

Flexible moulds will be used wherever required for admitting of undercuts in order to give the work a bold and crisp effect.

All pier caps, arches, panels and cornices of rotunda will be plaster, as designated on drawings.

CEMENT WORK AND SCAGLIOLA.

All bases, either plain or moulded, columns, pilasters, panels, the capitals of columns near stair rails in the Calhoun Street entrance vestibule, all window, door and panel architraves, jambs, reveals, cills and soffits throughout the Court House must be built up in cement and finished in pure gypsum scagliola, except where hereinafter specified to be of marble.

Architraves and jambs of all doors and windows, columns, pilasters and bases of same must be built up and finished in Keene Cement.

This cement scagliola is also to be used wherever denoted on the drawings either as cement or scagliola, and specified herein, generally consisting of all columns, pilasters and bases in the four Court Rooms, panels of walls and alcoves therein, all walls, pilasters and bases in judiciary, public and communicating corridors, witness corridors, private stair case hall, full run, pilasters, bases and panels in upper part of grand stair case hall, wall and door architraves between inner dome and the balcony around rotunda under inner dome, and wherever designated on drawings, also the door and window architraves in the office rooms, vestibule, and water closet of Power Station.

This work to be built up with a base of F F F New York plaster-of-Paris five-eights of an inch in thickness mixed with fine sharp sand in the proportions of two parts plaster to one of sand. The surface of finishing coat to be clear plaster three-eights of an inch in thickness.

All exposed surfaces to have a perfect stone polish; no artificial polish will be allowed.

All work to be executed with the very best materials and by the most practical artisans, adopting and using the very best practice for this kind of work.

The work to be solid color throughout the entire finish coat, with such tints or marble imitations as the Architect will select or designate.

All work and materials must be to the entire satisfaction of the Architect, and all persons employed for the execution of this work, must be subject to his approval or rejection.

MARBLE WORK.

All lavatories and water closets shown on plans for both Court House and Power Station to be wainscoted with marble seven feet high all around. The base to be ten inches in height, project one-quarter inch from the face of wainscoting, and rounded on the top corner. There will be a cap to all this wainscoting, six inches high, projecting one-quarter inch, with top and bottom corners rounded. The face of wainscoting to be flush with the face of plastering above. All of this marble to be selected white Italian. The plumbing specification calls for a portion of this work in connection with other marble work specified therein, and the contractor shall familiarize himself with that portion of the work and make the necessary provisions.

The water closet stalls to be four feet and nine inches deep, backs and sides one and one-eighth inches, and the fronts one and one-half inches thick, instead of as shown and figured on drawing No. 31, and all marble work shown on said drawing and specified in the plumbing to conform to the wainscoting herein specified.

The entrance vestibules, office rooms and machinery room of Power Station will have floor bases eight inches high, including the pit as well as the balcony of machinery room. All windows in machinery room, boiler room, water closets, and office rooms will have marble stools and aprons, the aprons to be four inches wide. All of this work to be pink Knoxville marble.

All floor bases throughout the Court House, besides those already specified for water closets, will be marble of the dimensions shown on drawings. Those on the Ground Floor will be Swanton Block, those on the Office Floor Vermont Verd Antique, and those on the Court Room and Attic Floors will be Champlain, excepting the four court rooms and alcoves for same, which will be Vermont Verd Antique. The floor bases of all columns will be marble corresponding with the other bases.

The bases of balustrades around well holes in rotunda, to correspond with the bases of their respective floors.

The balcony floors of Law Library, balcony over grand stair case and balcony under inner dome of rotunda will be Champlain marble.

The moulded bases and moulded base course throughout the Court House will be Champlain marble.

All windows throughout the Court House will have marble stools and aprons under same as shown on drawings. Those in the Ground Story will be Swanton Block, those in the Office Story will be Vermont Verd Antique, and those in the Court Room and Attic Story will be Champlain.

Throughout the vestibules, lobbies and rotunda of Ground Floor, all pilasters and piers between the moulded bases and plaster caps, and all walls between bases and plaster cornices, all stair and arch walls, groined ceilings, door and window jambs, reveals and soffits, except where otherwise specified, to be Champlain marble.

Throughout the vestibules, lobbies and rotunda of Ground Floor, all moulded architraves, jambs and soffits for same, window and tablet cills and corbels, to be English Veined Italian marble.

Throughout the vestibules, lobbies and rotunda of Office Floor, all pilasters and piers between the moulded bases and plaster caps, and all walls between bases and plaster cornices, all stair walls and groined ceilings, door and window jambs, reveals and soffits, except where otherwise specified, will be English Veined Italian marble.

Throughout the vestibules, lobbies and rotunda of Office Floor, all moulded architraves, jambs and soffits for same, window cills and corbels, will be blue veined Italian marble.

Wherever flush wall panels are shown, different shades of specified marble will be used.

All tablets will be white Italian marble.

Throughout the rotunda of Court Room Floor, the landing vestibules of grand stair case hall, around and opposite the elevators, the vestibule between Law Library and Tower, and the pilasters forming jambs of entrances to the judiciary and public corridors, all between the bases and plaster cornices and caps of piers and pilasters, will be blue veined Italian marble, except the following:

All work forming the ashler between said piers and pilasters on Court Room sides of rotunda and Law Library sides of vestibule, will be Istrian Fleuri or St. Baum.

The fluted pilasters on each side of the rotunda doors entering Court Rooms, and the vestibule door entering Law Library, will be English Veined Italian marble, the carved capitals of same to be white Italian.

The moulded architraves of these doors, also of the windows in rotunda between piers, and of the tablets between pilasters in vestibule, together with their jambs, soffits, cills and corbels, to be English Veined Italian marble. Said tablets to be white Italian.

The paneled architraves, carved consoles, panels, cornices and pediments of the rotunda and vestibule doors, together with the tablets over them, will be white Italian marble.

The stairs from Sheriff's Office to the Court Room floor, from Court Rooms No. Two and No. Three to their respective Jury Rooms in attic, and the stairs from Auditor's office to Ground Floor will have treads and risors of light pink or gray Tennessee marble. . Treads to be one and one-half inches thick. · Wall bases same as specified for floors.

The spiral stairs in Law Library will have treads only, one and one-half inches thick, to be of Champlain marble.

There will be marble risors and treads to all water closets throughout the Court House; the floor of closets to be six inches higher than the floor of rooms. These risors and treads will be the same marble as specified for their respective floor bases. Treads to be one and one-half inches thick.

All risors, treads and platforms for stairs in North and South Lobbies, West Vestibule and the stairs running from

the Ground Floor to Court Room Floor in East Vestibule,
to be Champlain marble. All treads and platforms to be
two inches thick, the treads to be in one length and the
platforms in one piece where possible, all thoroughly
secured to the metal work.

The bases and base course, baluster and baluster course
of balustrades and newels of said stairs and around well
holes in rotunda, to be Champlain marble, except the base
of balustrade around well hole in Office Floor of rotunda,
which will be as hereinbefore specified, and the bases and
dies of starting newels on Ground Floor, which will be Vermont Verd Antique, the carved cap members of same to be
white Italian marble. The two landing newels on Court
Room Floor will be perforated for illuminating purposes.

The caps and cap moulding, rails and rail moulding of
said balustrades and newels will be English Veined Italian.

The facias, and soffits of facias, under bases of all balustrades, to be English Veined Italian marble, including the
facias around well holes in rotunda and stair wells. All
floor nosings and other mouldings around wells that are
not shown or specified to be metal, will be Champlain marble.

Wherever marble comes in connection with plaster caps,
the neck moulding of such caps will be marble, not plaster.

The wall surface, from the large facia moulding running
around the grand stair case on Court Room Floor level,
down to the wall base of stairs, will be English Veined Italian marble, said facia moulding and the arch window sill
course above to be Champlain.

The flush wall balustrades will be the same marble as the
outer balustrades, and the flush wall panels to be formed of
contrasting shades or colors of marble selected by the Architect.

The contractor shall letter all tablets with such inscriptions as will be selected by the Commissioners and Architect. All letters to be sunk "V" shape.

All work specified herein and shown on the drawings,
including mullions, transoms, beams, floor nosings, caps,
bases, consoles, wall strings, panels and the like, must be
executed in strict conformity to detail drawings furnished
by the Architect.

All exposed surfaces, except carving, stair treads and
platforms, must be brought to a high and perfect stone pol-

ish; no artificial polish will be permitted. The carving to be unpolished, the finish to be as left by the carver. The stair treads and platforms to have honed surfaces, risors and nosings polished.

All carving and ornamental work to be executed from full sized models after designs furnished by the Architect. The contractor to furnish all models and submit them to the Architect for his approval before executing.

It is intended, and shall be demanded, that all marble shall be free from defects and from the best selected stock, of the highest grade and carefully selected for quality and color, and the beauty of the veining, to the satisfaction of the Commissioners and Architect.

Samples of all marble specified must be submitted to the Commissioners and Architect for their approval before ordering.

Pilasters, piers, walls, panels, etc., to be jointed substantially as shown on the drawings, in long lengths, and all work throughout must be so put up that the veins in the marble will match at the joints.

All jointing of the marble must be subject to the approval of the Architect.

The joints must be fine rubbed, set flush and close together without mortar, except bed joints, which will be in plaster-of-Paris. All to be close, square, exact and even, not to exceed one-sixteenth of an inch in thickness, and on completion to be pointed with plaster-of-Paris, colored to match the marble.

All external and internal angles will be mitred, the corners slightly rounded and highly polished. No joints to show on either face or sides except in the general bonding, and plumb bond must be maintained throughout.

Slabs are to be in no case less than seven-eights of an inch in thickness finished, and large slabs to be thicker to insure strength.

All marble work to be thoroughly secured in place with concealed clamps, dowels and anchors, all of brass. Drilling to be carefully done.

All work to be set in the most approved manner, and backed with the best Portland cement and mortar backing to insure a thoroughly first-class job throughout.

All joints, arises and mouldings to be perfectly and truly cut, all circular work true to radius, and in the case of groined ceilings, to be properly jointed and cut to conform to the work required.

The entire work to accurately conform to large scale and full size detail drawings to be furnished by the Architect.

The carving to be done by none but first-class carvers approved by the Architect, and any found incompetent shall be replaced by other carvers approved by the Architect and employed by the contractor.

MAIL CHUTE.

Do all work of preparing for and furnish and erect one mail chute at the point denoted on plans, extending from the Court Room Story to the Ground Floor, with inlets at each floor.

The chute to extend about four feet and six inches above the Court Room Floor, and to extend through necessary thimbles to mail box on Ground Floor.

Allow two hundred and fifty dollars for said mail box. All exposed metal work connected with this mail chute to be finished in copper deposit on iron, and the face to be of best plate glass.

The mail chute complete to be equal in every respect to those manufactured by Cutler's Manufacturing Co., Rochester, N. Y.

All instructions must be followed and the whole executed to the entire satisfaction of the Commissioners and Architect, and in accordance with the rules and regulations of the United States post office department governing the construction and arrangement of mail chutes.

CARPENTER AND JOINER WORK.

Provide a runway between the Jail and Power Station, entering the Power Station in the boiler room and entering the Jail at the nearest point in the present boiler room of the Jail. The runway to be built of two inch, thoroughly creosoted pine planking; all joints to be tarred when boxing is built. The cover of same to fit tightly over the top of trough, and to be fitted with strips running along the entire length and outside of the trough for holding cover in position; the cover extending at least one inch either side of the outside of trough. The inside measurement in clear to be fourteen inches wide by twelve inches deep. The contractor to do all excavating and filling necessary for proper placing the box underground.

All framing lumber and sheathing for roof of Power Station to be sound merchantable pine, thoroughly seasoned, free from sap, shakes, large or loose knots, or other defects that may impair strength, sawed square and of the dimensions required.

The roof to be constructed as shown on drawings, and sheathed with dressed, tongued and grooved boards, not over six inches wide, one and one-eighth inches thick after being dressed, laid close, through nailed at each bearing, butt joints on solid bearings. Grounds for fastening to be nailed to the sheathing where required, and tilting pieces and saddles to be provided.

The gutter beds to be formed of seven-eights inch dressed boards with graduating blocks underneath eighteen inches apart, properly graded from summit to outlets.

Provide all necessary and required lookouts for terra cotta cornices, securely built in place.

The iron work for roof is specified in structural iron work.

All lumber not otherwise specified that should enter the construction in any portion of the work to be best selected clear white pine.

The double hung sashes are shown by the double sash on drawings.

In both buildings all window frames and wood sills, including the outside moulding, to be selected clear white pine; the pulley stiles, together with the heads and parting strips, to be the best quality long leafed Southern pine. All parting strips to be fastened in place with countersunk head brass screws spaced not more than eighteen inches apart and set flush.

The frames to be constructed as shown on the drawings. Box frames to have extra long pockets in the pulley stiles for hanging sash, secured with countersunk head brass screws; within the pulley boxes metal strips to be hung to separate the weights.

All sills resting on masonry to be thoroughly bedded in hair mortar and as shown on details. The joints at sides and top between frames and masonry to be caulked tight with clean oakum. All frames must be securely fastened in place.

The frames must be well primed before being brought to the buildings.

To all box frames pulleys will be fitted and taken off until the sash are fitted. The pockets for pulleys to be properly protected during construction.

All outside sash to be two and one-quarter inches thick, and as shown on details; to be of selected clear white pine and veneered on the inside with thoroughly seasoned quarter sawed white oak, one-quarter inch thick.

All interior exposed surfaces to be quarter sawed white oak. All sash to be constructed in the most substantial manner, moulded, etc., as shown. The upper sash to have lugs at lower or meeting rail.

All double sash to be hung on best pulleys with steel ribbon sash line, sufficiently strong to properly hold and hang the sash; weights to be of compressed lead or iron as may be required. The hardware for this work to be furnished by the contractor and be of the very best and latest improved, and all to the entire satisfaction of the Commissioners and Architect. The hardware to be real bronze, plain finish.

All sash to be accurately fitted, hung true and run smoothly. The sash for Power Station to be made for A A double strength glass, and the sash for Court House to be made for plate glass one-quarter inch thick.

All to be provided for securing the glass in place with beads to be fastened with round headed bronze screws, The beads to be put in place with the other finish work.

All inside beads and finish, showing in rooms, to correspond in kind of wood to the finish of said rooms, otherwise all frames and sash to be as specified.

The exterior door frames for Power Station to be solid oak quarter sawed, two and one-quarter inches thick, and all doors veneered with quartered white oak.

The glazed partition between machinery and boiler rooms to be as indicated on drawings and of similar construction, design and detail as shown on details for Court House work; doors to be paneled and moulded to correspond with glass panels as shown.

All inside wood finish in Power Station to be quartered oak.

The screen partition to have door and side windows and

the large circular transom over to be as shown on details for windows in lobbys and rotunda of Court House.

All doors throughout the Court House to be as shown on the various drawings, built up of white pine core strips veneered with quartered white oak on faces and edges. The interior door frames may be either solid or veneered with white pine backing.

The water closet doors in lavatories will have moulding around panels, otherwise as shown on detail.

There will be a false door in alcove of assembly room corresponding with door on opposite side for closet.

Transoms wherever shown are to be pivoted and arranged to open easily, to be of the same thickness as doors below.

All grills shown in windows to be carefully framed up with moulded faces, etc., as shown, including large grills in arches of rotunda, Court Room Story.

All door frames, window frames, sash, jambs, doors, mouldings, caps, panels, etc., shown on the various drawings, marked on plans or called for in these specifications, and all other work consisting of the entire wood finish throughout both buildings, to be built up in the very best and workmanlike manner of thoroughly seasoned, best quality, selected, quarter sawed white oak, to show a good grain and be free from all imperfections. None but that with handsome grain to be used in the work.

All carving to be in the solid. The carving to be done with spirit, from full size models to be furnished by the contractor, made from designs furnished by the Architect. The models to be in clay or plaster and approved by the Architect before the work is commenced.

The entire work to be finished in the natural wood, and the work and material must be in accordance therewith, and to be left by the joiners perfectly clean, ready to receive a high cabinet finish.

The plain surfaces of all joinery work to be brought to a fine, even, smooth surface.

All mouldings to be accurately run; all wide plain surfaces, such as door panels, etc., to be double veneered on the best selected white pine core; to be, in case of doors and large panels, staved up.

All circular work to be built up solid and all framing, veneering, etc., to be thoroughly glued. All framing to be fitted and be allowed to stand and season as long as possible before gluing and wedging.

All panel work to be kept loose so as to allow for shrinkage. Where nailing through finished plain surfaces is required, it is to be chip-nailed, nails well set, and sliver glued in place, finished smooth and flush.

All wood work to be so worked and so put up that all nailing shall be secret, and the entire work to be done in best cabinet style and finish, and wherever possible the various mouldings, bases, caps, etc., to be tongued together.

All mouldings, etc., to be cut sharp and clean, joints and mitres to be neatly finished in strict accordance with drawings to be furnished.

The contractor must guarantee all panels in doors and all plain surfaces from cracking or warping.

No work to be put up until after the plastering and floors are finished and thoroughly dry.

All sash doors to have moulded beads for securing glass in place, to be put up with round headed bronze screws.

Double doors must be rebated and made to fit locks. All doors will be hung on three hinges each, and will have mortice locks and bolts; furnished by the county, but applied by the contractor.

At completion of the work all doors and sash must be in perfect working order and adjusted so as to work freely without unnecessary play.

HARDWARE.

The Commissioners are to purchase the finishing hardware for all doors, windows, transoms, closets, and water closets, comprising all finishing hardware.

This does not include any construction hardware such as sash weights, pulley ribbons, pulleys, bolts, nails, screws, anchors, ties, etc., all of which the contractor is to supply as herein specified.

The contractor is to furnish the information necessary to enable the Commissioners to supply the hardware above referred to at suitable times, and in satisfactory manner, and is to supply the information three months in advance of the time when the hardware will be required.

The contractor is to properly put in place all the hardware in the building, whether supplied by himself or the Commissioners, and shall do so in a first-class manner, and is to protect all, especially the highly finished pieces, such as escutcheons, door knobs, etc., from injury by workmen, and from being soiled, both while being applied and until completion of the buildings. All hardware, except that hereinbefore specified to be supplied by the Commissioners, to be supplied by the contractor, subject to the approval of the Commissioners and Architect.

All work must be executed to conform to the hardware furnished by the Commissioners.

GLASS AND GLAZING.

The Commissioners are to purchase all glass for windows, doors, transoms, ceiling lights and for partition between machinery room and boiler room of Power Station. This does not include the glass already specified for sky lights, clock dials, mirors, mail chute, etc., all of which the contractor is to supply as herein specified.

The contractor is to furnish the information necessary to enable the Commissioners to supply the glass above referred to at suitable times, and in satisfactory manner, and is to supply the information three months in advance of the time when the glass will be required.

The contractor is to properly secure in place all the glass in the building, whether supplied by himself or the Commissioners, and shall do so in a first-class manner, and is to protect all glass from injury until completion of the building.

All glass, except that hereinbefore specified to be supplied by the Commissioners to be suplied by the contractor, subject to the approval of the Commissioners and Architect.

The glass to be supplied by the Commissioners will be delivered by them on the premises for the respective buildings.

The contractor shall furnish the Commissioners with the sizes and shapes of all glass required, and be responsible for their correctness.

All glass in doors, windows and transoms to be secured in place with glazing beads fastened in place with round headed bronze screws as already specified, and all other glass to be secured in place as the nature of the work may require and to the satisfaction of the Commissioners and Architect.

All breakage to be made good by the contractor after

the glass is delivered to him by the Commissioners, and all glass to be clean and left in perfect condition at completion of the work.

PAINTING.

All iron work to be painted as specified, and all iron doors, floors, gates, railings around ceiling lights, iron stairs, balustrades, flag staffs, floor scuttle in lantern, ladders and all other such work to have two coats of pure white lead and best linseed oil paint, in addition to that already specified for iron work, tinted as directed.

The retaining cell to be finished in a third coat, varnished and tinted as directed.

The galvanized iron down spouts and cast iron pipes connecting the same, at Power Station, to be painted and sanded two coats in addition to priming.

All plumbing and exposed pipe work, not metal finished, to be painted two coats of best linseed oil and white lead paint of such color as the Architect may select.

All wood work which is to be painted to have a good coat of strong shellac before priming.

All exterior wood work to receive a priming, and three coats of best linseed oil and pure white lead paint, finished in such colors as will be directed. All priming of frames to be done before they are delivered to the buildings.

All paint to be properly prepared and laid on in good full body, well brushed in.

Pulley stiles, heads and parting strips of box window frames to be given two coats of linseed oil.

All door and window frames, sills, etc., against walls to have a good coat of yellow ochre and oil paint on the back; the outside of the window sash to be thoroughly primed before taken in the buildings.

The backs of all wood finish to have a good coat of pure white lead and oil paint, and all work to be filled and primed before it leaves the shop of the joiner.

None of this work to be taken into the buildings until the walls and floors are perfectly dry and the buildings enclosed.

All interior wood finish will be quartered white oak, to be properly prepared and brought to a fine cabinet finish; to be stained if so required, to have a coat of Wheeler's Patent Filler, or a filler equally good and satisfactory to

Commissioners and Architect, properly put on to fill the grain evenly, followed by one coat of white shellac, sandpapered to a smooth surface, and three coats of Berry Brothers Hard Oil Finish, or its equal, in the opinion of the Commissioners and Architect. Rub first coats with hair cloth or curled hair, and the last coat with pulverized pumice stone and raw linseed oil to a perfectly smooth dull gloss.

The outside doors and door frames of Power Station to be finished with best Exterior Wood Finish.

HEATING SYSTEM.

The building shall be warmed throughout by a single pipe low pressure system, except the Court Rooms, these to be heated by the forced blast hot air system.

The class of radiation to be used shall be direct-indirect, except as hereinafter specified.

It is proposed to utilize the exhaust steam from the several engines, pumps and other apparatus in the Power Station, supplementing the system when required, by live steam from the boilers, by means of a pressure reducing valve.

PIPING.—Commencing at the Power Station end of Tunnel, the contractor will furnish and make all necessary connections to main steam supply and return pipes, and to exhaust header. There will be run in the Tunnel one fourteen inch "light" wrought iron lap welded pipe; this pipe to be used as a heating main and to be carried from the Power Station through the Tunnel and connect with the three main steam heating pipes in basement of Court House. There will also be run a six inch return pipe, same to be connected to the steam heating mains in the basement of the Court House, and carried back through the Tunnel to the Power Station, where all connections will be made as set forth, such as may be required for contractor's "special system."

In case the sizes of the fourteen inch and six inch main steam and return pipes can be reduced, due to the installation of the "special system," the contractor in reducing the sizes of these pipes, will submit a statement of the sizes which he purposes to install for the Commissioners' and Architect's approval.

Suitable brackets or supports for the main steam and return pipes in the Tunnel and their placing is provided for in "Tunnel Specification." In placing these pipes the contractor will provide for the same to be installed so as to have a positive fall of at least one inch in each forty feet of run; the fall in each instance to be from the Court House toward the Power Station, so as to insure a return of all condensation. The main steam pipe at the Court House end to enter at the highest allowable point. At the Power Station and of these main steam and return pipes, the contractor will make allowance for the expansion in such pipes for their entire length, having first anchored his pipes firmly and securely at the Court House end of Tunnel in a manner subject to the approval of the Commissioners and Architect.

In making the Power Station expansion connections there will be provided four ells in each pipe, there beng two drops for each; one at the first turn in the Tunnel at street and the other directly at the entrance to Power Station, all as clearly shown in the drawings.

Suitable provision for draining the main steam supply pipe will be made by the contractor, the connections being such as his special system demands. The drips to be so connected as to properly and thoroughly drain the pipe without permitting any local circulation to take place, such connections to be made at least every fifty feet.

The fourteen inch straightaway valve for Court House main will be furnished as elsewhere specified, and is to be properly placed in the main steam pipe.

The center of heating main to run from point "C," shown on Tunnel drawings, to "J" at grade 54.39, and at this point "J" to drop to grade of 51.68, and thence to point "K" grade 50.50; from this point to point "R" at grade 49.14, and on to point "S" at grade 46.61, then to point "P," grade 45.94, where pipe drops, turns and runs to Power Station, as shown in detail drawings.

Starting at the northwest corner of the basement, the heating main will be divided and be carried around as shown on the drawings, each main making a complete loop of one-half the basement as shown, returning by a common return pipe to the tunnel entrance and there connected with the return main. Pipes to start at the Tunnel as near the ceiling as practicable, and to have a positive fall at least one-half inch in each twenty feet of run, and to maintain such positive fall throughout the complete loop. Provisions for expansion to be made by offsets as called for on the drawings.

The drawings clearly show the general distribution of steam heating mains in the basement, but their exact location relative to walls and ceilings is not figured; the contractor being required to determine such precise location so as to permit of draining and expansion as called for herein.

All riser pipes will be connected to the top of the steam supply main in the basement, and carried to the floor indicated in the Riser Diagram; all connections made by offsets to allow for expansion. All risers to be located in such position as shown on the drawings.

Whenever it is necessary to change the position of these vertical risers on account of varying thickness of wall or for any other purpose, all the offsets must be made by forty-five degree angles, the pipes, however, in all cases being carried as near the walls as practicable.

The contractor to make all connections from steam heating main in the basement to the steam coil of each hot water heating tank, as specified elsewhere. In this feed pipe there will be properly connected an automatic temperature regulating valve, as specified elsewhere. The contractor is to furnish and attach to the steam heating coil of each heater one automatic valve used in connection with the "special system" referred to hereafter.

The sizes of all pipes are given on the drawings and contractor will understand that no pipes of smaller diameter than those as specified will be allowed, unless the same can be reduced on account of the installation of his "special system" as called for hereafter and, if such reduction is made, it will be subject to the approval of the Architect and a schedule is to be submitted by the contractor, same to be approved by the Commisisoners and Architect before the work is commenced.

Wherever radiators are called for, connections for the same throughout to be made above the floor level.

All radiators to be conected to the risers by single feed connections; should the length of any of these pipes exceed four feet, then a pipe one size larger than that as called for by radiator opening, shall be put in. These connections all to have sufficient pitch to insure under all conditions, a return flow of the condensation against the current of steam entering the radiator.

All pipes to be so arranged as to permit of a full and free circulation of steam throughout the entire building, and to allow an easy return of condensation to the main return pipe at the Tunnel.

Contractor shall avoid the use of all fittings or connections which will admit of any lodgement of water, using eccentric fittings wherever the size of horizontal pipes are decreased.

All places where pipes pass through floors, ceilings or walls, same shall be fitted with insulating thimbles, having a diameter one-half inch greater than the pipe, so as to leave a one-quarter inch air space all around; same to be provided with plates secured in place; all to be independent of and free from pipe so as to permit a free contraction and expansion of steam pipe without injury to the plaster or other finish. All plates to be nickel plated and polished.

The whole work to present a neat and finished appearance when completed.

All pipes used to be standard full weight wrought iron pipes, with all sizes over one and one-quarter inch lap, welded throughout, with flanged fittings for sizes   three inches and over.

HEATING SURFACE.—The building is to be heated throughout, from the Ground to the Court Room Story inclusive, and the minimum amount of superficial radiating surface shall be twenty thousand nine hundred and fifty-seven feet exclusive of all mains and risers, and this amount is deemed sufficient to warm the various stories in the coldest weather, and not less than this amount will be accepted.

All radiators for direct-indirect service, unless otherwise specified, to be those known as the "Crescent" or "Italian" flue radiators as manufactured respectively by the Titusville Iron Company and the American Radiator Company, or other make of flue radiator equally good and acceptable to the Commissioners and Architect. All sections to be fitted with screw nipple connections.

All radiators for direct service where marked D 4 on the drawings, to be the "four column" Buffalo standard or other make equally good and acceptable to the Commissioners and Architect. For all other direct radiation the same type as used for the direct-indirect service, will be installed.

With each flue radiator for direct-indirect service, there shall be furnished a detachable box base, the design of same to be subject to the approval of the Commissioners and

Architect. For all direct-indirect radiators on the Ground floor, the air inlet will be underneath the radiator, while for all others the air will be brought in at back of radiator above the floor level.

The height of all radiators for the Ground and Office floors to be not over thirty-two inches and not less than thirty inches unless otherwise called for on the drawings, and each to have not less than five square feet of prime heating surface per section. Those for the Court Room floor to be not over thirty-eight inches in height and to have seven square feet of prime heating surface per section.

All radiators and risers throughout above the basement to be painted with at least two coats of paint or Japan of such color as the Commissioners and Architect may direct.

The contractor to make all galvanized iron duct connections from his radiator bases to the openings left in the masonry walls or floors as the case may be. The bottom of the ducts to be as near the floor line as possible where connecting with wall openings. The ducts leading to radiators located in the Jury Rooms on the Attic floor, to be carried through the partitions to the outside walls as shown on the drawings. All entrance air ducts to be securely and tightly connected to bases provided, all to present a neat and finished piece of work when completed.

The radiating surface will be distributed throughout substantially as shown on the drawings; but it is understood that the Commissioners and Architect shall have the right to determine the exact location of any and all mains, risers or radiators and that any changes in position shall be made before the work has been installed and shall not be the basis of an extra charge.

The contractor will provide a suitable tempering coil made up of at least seven hundred and twenty feet of one inch pipe; this coil to be installed and placed in position in the fresh air intake, leading to the two fans located in the basement; contractor making all connections from steam heating main in the basement to the tempering coil, besides all drip connections from same to connect with return pipes. The form of the coil to be subject to the approval of the Commissioners and Architect.

VALVES.—All radiator globe or angle valves to be rough bodied and nickel plated all over; to be the "Crane"

renewable metallic disc radiator valves, or other valves
equally good and acceptable to the Commissioners and
Architect. All valves to be provided with wood wheel.

The contractor will furnish the above specified valves for
only such radiators as are not controlled by a Thermostatic
valve, but will attach these latter valves to his radiators.

The following to govern the minimum size of valves allowable for radiators:

One inch to supply a minimum of twenty square feet.

One and one-quarter inch to supply a minimum of forty
square feet.

One and one-half inch to supply a minimum of sixty
square feet.

Two inches to supply a minimum of one hundred and
twenty square feet.

In case contractor can reduce the size of raditor valves on
account of the installation of his "special system" as called
for hereafter, he will submit a schedule showing wherein
such reduction may be made, this schedule being subject to
the approval of the Commissioners and Architect.

When radiators are installed, the contractor shall attach
to each radiator an ordinary air cock with a large opening
in same, for the purpose of facilitating the expulsion of all
dirt and grit from the system. After the system has been
in operation for a period of three weeks, the contractor is to
replace these air valves with a screw plug.

A straightway valve is to be placed in the basement on
each riser pipe and in pipes to Ground floor radiators.
Above the lowest valve on all risers, as well as in all steam
pipes leading directly from the heating mains to radiators
on the Ground floor there will be fitted a one-half inch drip
cock located in such a position that all these pipes may be
properly drained.

All valves used throughout, other than radiator valves to
be brass bodied straightway valves up to and including
three inch; above three inch, to be iron bodied.

SUPPORTS.—All horizontal pipes in the basement to be
substantially supported by expansion pipe hangers, placed
not further than eight feet apart. The riser pipes being
supported and firmly secured in position as near as may be
possible in the center of their lengths, so as to prevent, as far
as possible, any vibration, and at the same time allow a free
vertical expansion from the point of support; allowance for

expansion of all vertical pipes must be made in the connections leading to the radiator, while that for basement mains will be made where connected to risers.

Wherever the pipes are supported from any part of the iron construction there will be placed between the pipe and the hanger some suitable incumbustible material for the purpose of avoiding as much as possible the transmission of sound from pipes to the structure.

CUTTING.—Wherever it is necessary to do any cutting through floors, ceilings, or walls for the proper running of pipes, as called for, same will be done by the contractor. In every case, after the completion of such work, the building to be left in a condition satisfactory to the Commissioners and Architect.

PIPE COVERING.—All steam and return pipes in the basement, together with all valves and fittings, to be thoroughly covered with a one inch "Magnesia," "Asbestos Sponge" or "Nonpareil Cork" sectional covering. All covering to be carefully put in place, using brass bands; all seams to be perfectly tight; the whole to be painted with two coats of an air tight paint, dark brown in color. The entire work to present a neat and finished appearance when completed.

The contractor will also cover the main steam and return pipes leading from the Power Station to the Court House; these pipes will be covered with a two inch plastic covering, or such other covering as may be acceptable to the Commissioners and Architect. In placing this covering the contractor will leave at least a three-eighth inch air space all around either pipe, putting a spacer every ten nches in the length of the pipe so as to divide this air space up into sections ten inches long. Over this covering there will be tightly stretched a heavy canvas which will be securely fastened in place; the whole will then be painted with three coats of white air tight paint. None but the best workmanship and material will be used throughout in this work.

INDIRECT HEATERS.—The indirect heaters to be used in connection with the fan system will be furnished and installed as elsewhere specified; the contractor will make all connections from heating main to the two headers from which the different stacks are fed, placing in each branch a straightway valve to permit of independent control. Contractor will run all drip pipes from stacks, to the entrance of Tunnel and make the connections thereto.

GUARANTEES.—The contractor shall make, in connection with the "General Requirements," the following guarantees for the complete system:

That in the entire heating system a complete, uniform, continuous and noiseless circulation of steam will be established throughout the mains, to each and every radiator, stack or coil, with steam at atmospheric pressure.

That each and every radiator, stack or coil will be fully heated in all its parts with steam at atmospheric pressure.

That the circulation shall be completed fully and freely without any parts of the apparatus filling with water and without hammering or surging in any of the parts of the pipes, radiators, stacks or coils within thirty minutes after the time the main valves are opened, and that after the circulation is made, it will be maintained.

That the material and workmanship shall be the best throughout, and the entire system free from any mechanical defects, and he further agrees to replace and make good any parts going to make the entire system, which may prove defective within three years from the date of final acceptance of his work, without cost to the County.

No guarantee as to maintenance of temperature will be called for from the contractor.

SPECIAL SYSTEM.—The contractor will furnish and install complete ready for service, with double equipment of pumps and all the necessary apparatus, attachments, piping and accessories incident thereto, what is known as the "Webber system of steam heating," or the "Paul" system complete with double equipment throughout, or such other system as may be equally good and acceptable to the Commissioners and Architect, such system to be attached to the system of piping as called for in these specifications and the drawings forming part of same.

TEMPORARY CONNECTIONS.—It must be understood and agreed by the contractor that he includes in his contract account all labor and material necessary for making such temporary connections as may be required during the construction of the system and before its completion; such connections to cover all the heating surface, valves, pipes, fittings and accessories that may be required for temporarily furnishing heat in such portions of the building as may be desired, and at such times and to such extent as may be called for from time to time by the Commissioners and Architect, and in submitting his proposal, the contractor does so with the understanding that at such time as may be required, the Commissioners or their representatives may be allowed the free use of the apparatus for temporary purposes; the contractor to furnish a man to look after and operate the same during the temporary usage thereof, such party to have the complete installation under his charge, it being understood and agreed that in the event of any damage resulting through mismanagement or other circumstances the Commissioners or their representatives are not to be held responsible for same.

STEAM HEATING OF JAIL BUILDING.—In addition to what is called for in the preceding Steam Heating Specifications the contractor will be required to make connection to the present system of steam piping in the Jail building, and all connections to the "special system," running a four inch pipe and make the connections to the exhaust header in the Power Station, carrying this pipe underground, in a runway specified for elsewhere, to the Jail building, and make all connections to the present steam header in the boiler room of said building. The main steam pipe will be pitched from the Jail building, having sufficient fall to properly drain the pipe, this fall to be at least one inch in each forty feet of run; the two and one-half inch return pipe from the system to be connected at the Jail and to have sufficient positive fall toward the Power Station to return all condensation. There will be furnished and placed in both the steam and return pipes suitable straightway valves, one being located in each pipe in the Jail building and one in each pipe in the Power Station. Each of these pipes between the two buildings to be thoroughly and tightly covered with a plastic non-conducting covering, or such other covering as may be acceptable to the Commissioners and Architect. Covering to be at least two inches in thickness, and over this covering canvas will be tightly drawn, properly fastened and thoroughly painted with two coats of an air tight paint. In placing his covering on these pipes contractor will provide at least a three-eighth inch air space all around the pipe, air space to be divided into sections by closing the space at least every ten inches in the length of the pipe.

STEAM HEATING FOR POWER STATION.—The contractor will furnish all labor and material necessary for the heating of the Power Station, all is outlined herein, the system of piping to be connected to the "special system" as referred to in the preceding part of these specifications.

All piping will be run substantially as shown on the drawings and will be of the sizes as indicated, unless the contractor can make a reduction in such sizes on account of the installation of his "special system," but before such reduction is made he will submit a schedule stating specifically where such changes may be made; this schedule to be subject to the approval of the Commissioners and Architect.

All radiators are to be connected by single feed connections.

All pipes are to be run and all fittings and connections will be made the same as outlined in the main body of these specifications. All insulating thimbles and plates will conform to those called for to be installed in the Court House.

The building is to be heated throughout by direct radiation. All radiators used to be thirty-two inches in height, unless otherwise called for on the drawing, and to be three column "Buffalo standard" radiators or other make of radiator, equally good and acceptable to the Commissioners and Architect. The total amount of radiation being not less than seven hundred and sixty feet.

The contractor will thoroughly paint all radiators and such pipes as are not required to be covered, with the best quality aluminum paint. The radiating surface will be distributed throughout the building substantially as shown on the drawings, but it is understood that the same conditions governing changes in Court House radiation will apply to this work.

All valves as used throughout to be the same as those called for to be installed in the Court House.

All horizontal pipes to be substantially supported by expansion pipe hangers placed not more than eight feet apart.

The contractor will thoroughly cover all horizontal pipes with a one inch "Magnesia," "Asbestos Sponge," or "Nonpareil Cork" sectional covering; all covering to be carefully put in place, using brass bands, all seams to be perfectly tight; the whole to be painted with at least two coats of a white, air tight paint; the entire work to present a neat and finished appearance when completed.

The contractor will make guarantees to cover the work under this heading, same as are called for in the main body of these specifications, particularly referring to the Court House heating.

TEMPERATURE REGULATION.

These specifications are intended to cover the furnishing and installing of a complete and perfect system of automatic heat regulation and includes thermostats, thermostatic valves, mixing dampers with necessary frames, air compressor and tanks together with all other accessories going to make a comprehensive system covering all the work as outlined hereinafter under this heading.

The "Johnson" or other system of automatic temperature regulation equally good and acceptable to the Commissioners and Architect will be allowed.

The contractor will furnish all labor and material neccessary to perform the work for a complete system as called for, and to facilitate the completion of his contract, he will familiarize himself with the Steam Heating and Hot Blast system specifications, as this work forms part of each of these other completed systems.

The intention is to maintain a constant temperature of seventy degrees Fahrenheit in all the rooms to be controlled and the thermostats will be so designed as to operate the controlling devices within a variation in temperature of one degree above or below this point.

The location of all thermostats placed in the rooms will be determined by the contractor, but it is understood that the Commissioners and Architect reserve the right to change the location of any or all of such, provided the same does not affect the operation of the system and is changed before work has commenced, and that any change so made shall not constitute the basis of an extra charge.

THERMOSTATS, VALVES AND DAMPERS.— Thermostats are to be placed in each and every room where same are called for on the drawings. In all rooms other than the four Court Rooms, the thermostats will operate the radiator valves furnished (to the number as indicated on the drawings,) by the contractor and attached to such radiators as required by the Steam Heating specifications; the sizes of radiator valves to be in accordance with the tabulation given for same under "Steam Heating Specifications;" these sizes, however, to be modified should the "special system" installed by the Steam Heating contractor permit of same being reduced, such reduction being subject to the approval of the Commissioners and Architect.

The Court Rooms will be heated by the Hot Blast system, each room having four separate hot air ducts, two of those for each room being led from the same mixed air chamber in the basement. To control the mixing dampers, there will be installed in each Court Room two thermostats so designed as to operate their respective dampers by a graduated motion, mixing the hot and cold air in such proportion as to maintain a temperature of approximately seventy degrees in that portion of the room in which the thermostat is placed. The contractor will provide the necessary dampers which will be so constructed as not to appreciably reduce the cross sectional area of ducts or hinder the free movement of the air irrespective of what position the dampers may assume. The heated air will be admitted until the temperature of the room shall have reached sixty-eight degrees, and at such time the dampers will then take an intermediate position, maintaining a temperature of seventy degrees in the room.

FINISH.—The finish of all thermostats to be what is known as "Bower Barff." All thermostatic valves to be rough body, nickel plated all over, the diaphragms to be finished with dead black paint to correspond with the thermostats, the rims to be nickel plated and polished; valves to be provided with renewable metallic discs and nipple connections. All air pipes to be concealed as far as the finish of the building will permit.

AUTOMATIC AIR COMPRESSOR.—The air compressor and tank for providing the necessary air for the operation of the valves, dampers, etc., to be furnished and installed at some convenient point in the basement, and placed near the other machinery and apparatus there located. The air compressor will be of such capacity as the system demands, and to be operated by water pressure. The pressure to be supplied by two independent sources, viz: the house tank and the city mains. The contractor to make all necessary water connections from basement mains, of galvanized iron pipe and provide brass bodied straightway valves so as to permit of water being supplied from either source independently of the other; valves to be located adjacent to the compressor. The discharge pipe to be carried to the catch basin.

Contractor to make all necessary connections to the system and furnish proper and substantial supports for his apparatus, all subject to the approval of the Commissioners and Architect.

AUTOMATIC HOT WATER REGULATORS.—There will be furnished and installed two thermostats for controlling two automatic valves to be furnished by the contractor and attached to the steam coil of each hot water heating tank as required by the Steam Heating specifications. All connections to be made complete; the devices being so arranged as to automatically maintain the temperature of the water in the tank at one hundred and eighty-five degrees; the valves to be operated by air pressure same as all radiator valves provided by the contractor.

CUTTING AND FITTING.—Wherever it is necessary to do any cutting through floors, ceilings or partitions for any purpose, the same shall be done by the contractor. In every case after the completion of such work, the building to be left in a condition satisfactory to the Commissioners and Architect.

GUARANTEES.—Contractor will guarantee, in connection with the "General Requirements," the perfect working of his entire system mechanically. That a temperature varying one degree either above or below seventy degrees will be maintaind in all rooms to be controlled, provided the supply of steam to the radiators is maintained, (steam being supplied at atmospheric pressure.)

That all material and workmanship throughout will be of the best of its respective kind that can be obtained and to replace and make good any defects which may develop due to defective workmanship or material within three years from date of final acceptance of the work.

FORCED BLAST HEATING SYSTEM.

GENERAL.—The following specifications are intended to govern the furnishing and installing of a complete Forced Blast System for heating the four Court Rooms.

This system is to be entirely separate and distinct from that as used for heating other portions of the building, and contemplates the furnishing and installing of fans, heating coils, housing, electric motors and ducts, all complete and ready for operation as hereinafter set forth.

The contractor will familiarize himself with the other heating specifications, which work is intimately connected with the successful operation of this, for the purpose of facilitating the advancement of the work.

FANS.—There will be installed in the basement, two, three-quarter housing, top horizontal discharge, steel plate fans, with double heaters, arranged to blow through heaters with cold air by-pass underneath. Each fan to have a capacity of delivering at least seven thousand seven hundred and sixty-five cubic feet of heated air per minute to Court Rooms through the system of ducts as shown on the drawings. Besides this, each fan will be capable of delivering at least two thousand eight hundred and fifty-eight cubic feet of tempered air; this fresh air to be carried through the system of ducts as shown to the several rooms on the Ground Floor, entering box bases beneath the radiators in such rooms. The total capacity of each fan will therefore be at least ten thousand six hundred and twenty-five cubic feet of air discharged per minute.

The speed of these fans to be not over one hundred and fifty revolutions per minute. Fans to be connected by means of an improved friction clutch so arranged as to permit of either fan being operated independently of or in connection with the remaining fan. The complete apparatus to be securely set upon substantial brick foundations furnished in connection with this work.

The completed housing and intake chamber to be made of number fourteen iron plate properly riveted and stiffened with angle irons.

These fans to be one-hundred inch steel plate "Buffalo" "Sturtevant" or other make of steel plate fan equally good and acceptable to the Commissioners and Architect.

In addition to the above fans, there will be two seventy-inch steel plate exhaust fans installed in the Attic. The make of fans to be "Buffalo" or other make of fan equally good and acceptable to the Commissioners and Architect. The speed of the these fans not to exceed two hundred revolutions per minute. Fans to be provided with overhung wheels so as to permit of belts being attached in each instance entirely outside of the exhaust chamber. The purpose of the fans is to exhaust the foul air from Court Rooms

Numbers One and Two, and Three and Four respectively, each fan having a capacity of exhausting at least five thousand five hundred cubic feet of air per minute through the system of ducts as shown on the drawings. Fans to be set substantially as shown on the drawings; exhaust chambers for the reception of fans to be provided as elsewhere specified; the contractor to see that the same are made tight and suitable for the reception of this apparatus; all supports for fans to be provided by contractor; those necessary for motors to be provided for, as elsewhere specified and laid out in accordance with templates to be furnished upon signing of the contract.

All fans to be provided with self-oiling bearings; to be perfectly balanced; to run noiselessly and without vibration; it being absolutely necessary that all machinery operate quietly so as to prevent noise being communicated to the Court Rooms.

Contractor will state in his proposition the horse-power required to drive each of the fans which he is to furnish under his contract.

MOTORS.—With the fans as installed in the basement, there will be furnished two electric motors, each of sufficient capacity to drive its respective fan without overloading. Motors to be belted to pulleys of fans. Motors to be securely installed upon suitable concrete foundations provided by the contractor.

For driving the exhaust fans in the Attic, there will be provided and installed with each fan an electric motor of ample capacity to operate its respective fan without overloading. Motors to be belted to fan pulleys, the pulleys being located outside of exhaust chamber.

All motors to be provided with base frames and adjustable belt tighteners. With each motor as proposed, contractor will provide a suitable endless double leather belt made of positive center cuts and strictly short lap. Belting to run true and noiselessly.

With each motor there will be provided a "Whittingham" solenoid automatic motor starter of the latest and most approved design, and a double pole snap switch.

There will also be provided with each motor, an approved non-combustible speed regulator. All to be properly installed and connected.

Motors to be perfectly balanced and to operate without

vibration and to be installed in the most approved manner.

SECTIONAL HEATERS.—The two sectional heaters to be provided with the basement fans will contain at least six hundred square feet of heating surface each. Each heater to be made up of three independent sections of two hundred square feet, having a separate connection to the steam header or each section, as well as separate drip connection to return pipe. In each feed and drip pipe, there will be placed a brass bodied straightway valve. Provision is made elsewhere for all connections to heaters and drips and such valves as that part of the specification calls for, the contractor shall however provide all flanged connections for headers complete with necessary bolts, nuts and "Rainbow" or "McKim" gaskets.

Header, individual feed and drip pipes to be of ample size to care for the surface supplied or drained, the steam being furnished at atmospheric pressure. Each heater shall be thoroughly tested before shipment, and after installed must be perfectly tight and free from leaks. Heaters to be made of wrought iron pipes screwed into cast iron; to be so designed as to permit of free expansion without causing leaks or damage to any part of the complete heater, and to be equal to those as manufactured by the Buffalo Forge or Sturtevant companies.

ARRANGEMENT.—The fans in the basement will be so arranged as to take fresh air into the chamber located between the two and to blow through their respective heaters, with a cold air by-pass underneath. Fans will be right and left hand and heater section connections will be made on the outside of the chamber which enclose entirely the fans and heaters.

Flap dampers will be arranged so as to permit of the discharge through ducts leading to Court Rooms, being carried from one fan to all ducts or from each individual fan to its own respective ducts at will. Dampers will be placed in the cold air by-pass.

DISTRIBUTING DUCTS, ETC.—The fresh air will be brought down from the roof of the building and be carried to the fan chamber as shown on the drawings, the contractor providing and installing all the necessary galvanized iron ducts for this intake as well for all double ducts throughout from fan chamber to mixing chambers and thence to the separate Court Rooms, all as shown on the drawings. In

addition to these ducts, there will be cold air ducts running
around the basement, with outlets into radiator bases as
shown. In these ducts flap dampers will be placed at each
and every point where a branch duct is connected with main
supply duct, these dampers to be so arranged as to permit
of their being permanently fastened after having been finally
adjusted by the contractor, so as to admit the correct
volume of air beneath radiators.

All ducts throughout to be as shown on the drawings;
rectangular pipes or ducts to have all flat surfaces well stayed
and braced to prevent any buckling and all vibration; all
joints must be well riveted together, and soldered so as to be
absolutely tight. Where main pipe is increased in size on
account of branch ducts entering, such increases will be
gradual, the duct flaring to such an extent that the angle
made by the flaring with the line of main duct will in no
case exceed fifteen degrees.

The following will govern the guage of iron to be used in
all cases for circular ducts or flues:

Up to and including eight inches diameter, number
twenty-six.

Nine inches to eighteen inches diameter, number twenty-
four.

Nineteen inches to thirty-six inches diameter, number
twenty-two.

For rectangular ducts:

Maximum dimension eight inches and under, number
twenty-six.

Nine inches to fourteen inches, number twenty-four.

Fifteen inches to twenty-two inches, number twenty-two.

Twenty-three inches to thirty-six inches, number twenty.

Thirty-seven inches and larger, number eighteen.

All ducts to be done in a thorough and workmanlike manner, the whole system of piping securely fastened and held
in place with approved wrought iron hangers, brackets or
posts, and wherever suspended from the iron structure of
the building, an elastic separator made of some incombustible material will be placed between the hanger and the
duct or ducts so supported, for the purpose of preventing
transmission of sound. The fresh air intake to fan chamber
will be carried just above the floor line of basement.

Where duct connections are made through the floor with

the radiator box bases, the contractor will see to the ducts being secured firmly in position, having all joints tight.

The exhaust chambers in the attic will be furnished as elsewhere specified, complete with fan supports or foundations laid out from the contractor's templates. The motor sliding bases to be securely fastened to the flooring or other framing, by the contractor. The contractor must see that the chambers are made tight and adaptable to the reception of the apparatus which he is to furnish. Exhaust heads and all intermediate connections between chambers and the roof to be provided and in position prior to the completion of the roof of the building; these discharge heads to be so designed and arranged as to permit of a free and easy discharge for the air and at the same time afford a positive protection from the elements; heads to be provided with adjustable dampers so that opening to atmosphere may be partially or entirely closed if necessary or convenient. The contractor will also provide and properly place a head over the fresh air intake, leading to basement; opening to be screened with brass wire cloth twelve mesh to the inch and of number twenty-four wire; head to be so designed as to prevent the entrance of rain or snow and at the same time offer no obstruction to the free and easy entrance of air. All heads, whether for exhaust or intakes to be made of number twenty cold rolled copper and to be securely fastened in position.

GUARANTEES.—The contractor guarantees, in connection with the "General Requirements," the entire installation which he installs, to be the latest and most modern in all particulars, none but the very best workmanship and material being used throughout and he further guarantees to replace and make good any defects which may develop, under proper and normal conditions of operation, due to faulty workmanship or material within one year from date of final acceptance of his work, without cost to the County.

That the apparatus which he furnishes will perform the work for which it is intended with the conditions as herein stated, the intention being to maintain a temperature of seventy degrees Fahrenheit in each and all Court Rooms.

VENTILATION.

GENERAL.—The following specifications govern proposals for furnishings, setting up in place and making all connections as may be required for a complete and comprehensive system of ventilation, as hereinafter set forth and specified.

SYSTEM.—The system of ventilation to be employed throughout will be the exhaust type mechanical ventilation.

ARRANGEMENT.—The several exhaust fans as herein specified will be located in the Attic substantially as shown on the drawings; the exhaust chambers being provided as elsewhere specified; the contractor to see that same are made tight and suitable for the reception of his fans; foundations being prepared as elsewhere specified and laid out in accordance with the templates to be furnished upon signing of the contract.

All fans will discharge vertically through the roof; heads and all intermediate connections between chambers and the roof to be provided by the contractor, same to be placed in position prior to the completion of the roof of the building; heads to be made of number twenty cold rolled copper and to be so designed and arranged as to permit of a free and easy discharge of air and afford a positive protection from the elements; adjustable dampers to be provided for partially or entirely closing the opening to atmosphere, these dampers to be so designed as to in no way obstruct the free discharge of air.

FANS.—The contractor will furnish and install in position in each of their respective exhaust chambers, the fans and make all connections complete, as specified. There will be five steel plate, full housing, up discharge, belt driven fans of the "Buffalo," "Sturtevant" or other make equally good and acceptable to the Commissioners and Architect. Fans to have overhanging wheels and to be provided with self-oiling bearings. Each fan to be complete in all details ready for belting to its respective motor.

The fan to be placed in chamber marked "A" on the drawings, to be capable of exhausting not less than four thousand cubic feet of air per minute through ducts as shown, from all rooms connected to this chamber, discharging same into the atmosphere.

Each of the fans to be located in exhaust chambers marked "B"-"E" on the drawings, will be capable of exhausting not less than two thousand four hundred cubic feet of air per minute through ducts as shown, from all rooms connected to such chambers, and discharge same through the exhaust head to atmosphere.

The fans to be placed in the exhaust chambers marked "C"-"D" on the drawing, will each be capable of exhausting not less than three thousand cubic feet of air per minute through ducts as shown, from all rooms connected to these chambers, discharging same to atmosphere.

The speed of any fan not to exceed two hundred revolutions per minute.

All fans furnished under these specifications to be perfectly balanced; of the best workmanship and material throughout and to run noiselessly and without vibration. Noise must absolutely be reduced to a minimum on account of the close proximity of the Court Rooms.

MOTORS.—With each exhaust fan proposed there will be furnished an electric motor of sufficient capacity to drive its respective fan without overloading. Each motor to be provided with sliding base and adjustable belt tightener; to be securely fastened to the flooring or supports; to have approved self oiling bearings; to be mechanically and electrically perfect, with armature perfectly balanced and to be of the latest and most approved design.

With each motor, as proposed, the contractor will provide a suitable endless, double leather belt made of positive center cuts, strictly short lap. Belts to be used to drive fans. Belting must run true and noiselessly. There will also be furnished with each motor one "Whittingham" Solenoid automatic motor starter of the latest and most approved design, and one approved double-pole snap switch.

DUCTS.—The contractor to provide all galvanized iron branch and main ducts, and make all connections to vacuum chambers, substantially as shown on the drawings. Where main ducts enter exhaust.chambers they will be suitably flared to reduce the velocity of incoming air. Where each main duct enters an exhaust chamber, there will be provided a suitable damper, arranged so as to partially or entirely shut off all communication with the chamber. Each and every branch or pipe must be provided with a regulating damper. All ducts and flues to be well riveted and soldered, all joints to be left tight and in a finished condition.

The guage of iron to be used in all ventilating duct work throughout, to be goverened by the tabulation given under the specifications for "Forced Blast Heating System."

The sizes of all flues, main and branch ducts are given on

the drawings and these sizes will be strictly adhered to by the contractor.

REGISTERS.—All registers called for on the drawings, including those for the Court Rooms, to be furnished and put in position by the contractor. Wherever two vents are called for, the intention is to locate one as near the ceiling as the finished cornice will permit, and the other near the floor line. All registers to have wings or louvers, those placed near the ceiling to be controlled by cord pulls conveniently arranged for operating, provided with Lignum vitae or Celluloid pulls on ends.

All registers unless otherwise specified to be white Japan finish and of design similar to Tuttle and Bailey's "Empire."

Contractor to furnish all labor and material for setting these registers firmly and in a neat substantial manner; connecting them to their respective boxes or ducts.

The inlet registers for the four Court Rooms to be electro-bronze plated; the design such as the Commissioners and Architect may select. The outlet registers to be located around the skylights as indicated on the drawings for each separate room.

On the drawings where vents are called for to be placed in closet doors, they will be eight inch by ten inch ventilating plates unless otherwise stated. Plates in all cases to be electro-bronze plated and of a design such as Tuttle & Bailey's "Empire," or other design equally good and acceptable to the Commissioners and Architect. These plates will be securely fastened in place in the lower panels of their respective doors, all to the satisfaction of the Commissioners and Architect.

VENTILATOR.—In addition to the exhaust heads and registers as specified, the contractor shall furnish and install one Combination Skylight "Star" ventilator, twenty-four inches in diameter. Ventilator to be made of number twenty cold rolled copper; to be placed in position on the roof of the building, and fitted to the skylight directly over Clerk's and Sheriff's stairway. Contractor will furnish and securely fasten in place white Japaned open ventilating plates four inches in width, extending entirely around the border of ceiling light; providing all intermediate connections between the ceiling light of well and the roof; this connection to enclose the entire ceiling light and vents and to extend up to and enclose the entire skylight in the roof. In the ventilator, contractor will provide a suitable damper with chains attached, same to be carried down between the wall over suitably arranged pulleys or trolleys and connected so as to be operated from the landing of stairway; the operating mechanism to be in the form of an ornamental bronze indicator or dial marked "open" and "closed." Ventilator to be securely fastened in position. The work to present a neat and finished appearance when completed.

CUTTING AND FITTING.—Special attention is called to the cutting of holes in floors, ceilings or walls for the running of ventilating pipes. Wherever cutting is necessary, same will be done at the expense of the contractor, who will in all cases obtain the consent of the Commissioners and Architect or the Building Superintendent before any such cutting is done. After the completion of his work, the contractor must leave the building in a condition satisfactory to the Commissioners and Architect.

CONCLUSION.—The contractor must familiarize himself with the other work that may in any way be dependent upon or connected with the ventilating work.

All work included under this specification to be completed in all particulars; the entire ventilating system being thoroughly modern, complete and comprehensive throughout; the contractor providing all labor and material necessary to perform the work in accordance with the true intent and meaning of these specifications, although each and every item necessarily involved in the work is not specifically mentioned.

GUARANTEES.—The contractor guarantees, in connection with the "General Requirements," the completed ventilating system to perform the work for which it is intended.

That only the best workmanship and material will be used throughout; and to replace and make good any defects which may develop due to faulty workmanship or material within one year from date of final acceptance of the work, without cost to the County.

That his motors and exhaust fans will operate quietly and without vibration.

SEWAGE AND PLUMBING.

IN GENERAL.—The contractor shall furnish all labor and material and build and construct in a good, firm and sub-

stantial manner, in place in and around the building, the
sewers, man-holes, catch basins and their appurtenances, and
all piping for water supply, wastes, ventilation of wastes and
the plumbing fixtures and all appurtenances, making the
system of water supply use and waste complete and ready
for service, as shown upon the plan and set forth in the following specifications.

The Power Station to be included in this work.

MATERIAL.—All material used in this work shall be the
best of its kind.   All pipe and fittings shall be of the inside
diameter designated.   Cast iron pipe shall be extra heavy
soil pipe, each pipe marked and bearing the maker's name,
of equal thickness throughout and of proper weight and
coated inside and outside, while hot, with coal tar varnish.

Fittings for cast iron pipe shall be specially made for the
purpose for which they are used, of equal thickness
and weight, and of the same class: and of the same inside
diameter and coated the same as the pipe with which they
are used.   Fittings for hand-holes, closed with a trap screw,
shall be used where shown or specified.

Wrought iron pipe shall be standard and all sizes of greater diameter than one and quarter inch, shall be lap welded.
When used for waste or ventilating pipe, it shall be coated
inside and outside, while hot, with coal tar varnish.   When
used for water supply, it shall be galvanized pipe.

Fittings for wrought iron pipe for soil, waste or ventilating pipe shall be of cast iron, specially made for this purpose,
of same inside diameter as the pipe with which they are
used, and those supporting risers shall have a proper shoe
cast on them.   Graded fittings shall have the grade cast in
them.   Fittings for water pipe shall be of malleable iron and
all changes in the direction of water pipe and connections
to pipe shall be made with water elbows and water tees; all
fittings shall be coated or galvanized, as specified for pipe
with which they are used.   All wrought iron pipe or fittings
shall be cut with full threads and when used for waste or
ventilating pipes, cut to a guage.   Fittings for junctions of
cast iron and wrought pipe shall be cut at one end with full
threads.

Lead pipe for water supply and supply connections shall
be strong pipe, weighing as follows:

One-half inch pipe, one pound, twelve ounces per lineal
foot.

Five-eights inch pipe, two pounds eight ounces per lineal foot.

Three-quarter inch pipe, three pounds per lineal foot.

One inch pipe, four pounds per lineal foot.

For waste and ventilating connections, light lead pipe shall be used, weighing as follows:

One and one-quarter inch pipe, three pounds per lineal foot.

One and one-half inch pipe, four pounds per lineal foot.

Two inch pipe, five pounds per lineal foot.

All exposed pipes or fittings for connections in toilet rooms shall be of polished brass, nickel plated.

Sheet lead for safing and flashing shall weigh three pounds per square foot.

All brackets, clamps, pipe hooks, or hangers shall be of wrought iron and constructed to the satisfaction of the Commissioners and Architect. Clay pipe shall be hard burned or vitrified, salt glazed, socket pipe, equal in quality to "Standard" Akron pipe.

Brick shall be hard burned sewer brick, free from lime, and shall be thoroughly wet before laying.

Portland cement shall be best German Portland, approved by Architect. Sand shall be clean and sharp. Mortar for brick work shall be made of one part Portland cement and two parts sand; for pipe laying, one part Portland cement and one part sand,—all by measure. The cement and sand shall be mixed dry to a uniform color, and then wet with as little water as will make mortar of proper consistency when well worked. All mortar which shall have set or become hard in the box, shall be thrown out and shall not be used in the work.

Lead shall be soft pig; gaskin shall be picked oakum.

The contractor shall do all excavating necessary in laying in the work included in this specification, and shall furnish all necessary sheeting and bracing and shall use the same for the protection of walls and foundations wherever necessary, or when ordered by the Architect or Superintendent, and shall keep the trenches free from water during the progress of the work by pumping or bailing.

A line shall be used to mark out sewer trenches, and there shall be no variation from the plan, except on order of the Architect or Superintendent.

For water pipe, the trenches outside of building shall not be less than six feet deep below grade.

SEWERS.

CLAY PIPE SEWERS.—Commencing at street sewer where shown, the contractor shall connect therewith with vitrified pipe and run the same as shown on plan to a point two feet outside of building wall. Each pipe shall be laid in line, rising at the rate of one-eighth inch to one foot of length and firmly bedded. Each joint shall be tightly caulked with Portland cement mortar and mortar thrown on and struck off smooth outside to the proper bevel. All joints shall be cleaned out inside. Dry sand shall be filled in and tamped solid to six inches above pipes. Above this, the trenches shall be back filled as hereinafter specified. All existing street connections, if any, shall be removed to a point not less than one foot outside of building and blocked off with brick and cement mortar. The connection of iron sewer to clay sewer shall be made with a collar of brick laid in Portland cement of two rings of brick with not less than six inches bearing on each side of joint.

CAST IRON SEWERS.—All sewers laid in ground through and inside of building shall be of cast iron. The pipes shall be bedded their entire length in dry earth or sand with a proper bell hole at each joint. All pipes shall be laid truly in line and grade with a grade of one-quarter inch to one foot of length, unless otherwise directed by the Architect or Superintendent. Any change in line or grade shall be made with proper curve fitting, and any junction shall be made with a Y branch.

Hand hole fittings shall be used when shown and each hand hole shall be fitted with a brass trap screw, leaded and caulked as specified for joints. Fittings receiving risers shall be firmly supported. All joints shall be tightly caulked with hemp gasket, leaving not less than one and one-half inch depth of lead room and then thoroughly filled with molten lead and caulked tight. The lead shall be cut off smooth outside and care shall be taken that the joint is left perfectly smooth inside. All openings for connections shall be closed with plugs until used, and all hand holes shall be closed at once, and the sewer shall be kept clean. Any dirt which shall get in shall be removed by the contractor.

BACK FILLING.—As fast as pipe is laid, dirt shall be rammed in place at the sides of the pipe, leaving the joints and tops exposed until tested, as hereinafter specified.

After testing, the trenches shall be filled in layers not more than nine inches deep, each thoroughly rammed.

SUSPENDED SEWERS.—All suspended sewers shall be of wrought iron pipe suspended to the ceiling of basement as near the ceiling as possible, rising one-quarter inch to one foot of length unless otherwise directed by the Architect or Superintendent. The pipes shall be firmly secured to beams with wrought iron hangers, not more than twelve feet center to center, and each fitting for riser shall be seated upon a bracket or proper hanger, bolted or clamped securely to wall or beam. All junctions shall be made with a Y branch, and each deflection in line or grade shall be made with proper curved fittings. Each Y shall be provided with a proper boss or hand hole closed with a plug. Openings for hand holes shall be full diameter of pipes. All joints in and to wrought iron pipes and fittings shall be made with threads fully coated with red lead and oil.

TESTING.—At such time as Architect or Superintendent shall direct, the contractor shall stop all openings in the cast iron sewers and shall fill them with water. If any leak shall show in the pipe or joint, he shall make the same good and repeat the test, and shall not be entitled to demand or receive payment until the same is made tight, as contemplated herein. After this test, the back filling shall be completed as before specified.

CATCH BASIN.—Where shown, the contractor shall furnish, set and connect J. J. Wade's sectional iron catch basins, figure eight, Wade's catalogue: basin to be thirty inches diameter and thirty-four inches deep below outlet, and made perfectly tight. Inlet of basin shall be connected to drips of tanks and pumps and outlets of basin, to sewer; the flushing connection of basin to be plugged.

MAN HOLES.—At running trap, where shown, the contractor shall build circular man hole. Man hole shall have two flat courses of brick in the bottom and two rings of bats in wall, laid in full joints of cement mortar, plastered outside and cleaned and pointed inside. Man holes shall be drawn in at top and covered with a dressed flag stone with countersunk iron ring and screw flush lid, eighteen inches in diameter.

REMOVAL OF EARTH.—After back filling as before specified, the contractor shall remove from the building all

surplus earth resulting from his work and dispose of same
at his own cost and expense.

PLUMBING.

JOINTS.—Joints in and to cast iron pipes and fittings
shall be made with caulked lead as specified for sewers,
using on lead connection a heavy cast brass ferrule.

Joints in and to wrought iron pipes and fittings shall be
made with threads fully coated with red lead and oil and
screwed perfectly tight. Joints in lead pipe or of lead pipe
to brass fitting, shall be of solder neatly wiped. Joints of
lead pipe to wrought iron pipe shall be made with a brass
ferrule. All joints must be finished smooth inside.

CARPENTER WORK.—The contractor shall do all
cutting and fitting necessary in securing his pipe, and all
carpenter work in connection with the troughs, gutters,
and safes hereinafter specified, and all necessary for ground
supports, etc., in setting the fixtures and marble included
in this specification.

RISERS AND BRANCHES.—All risers and concealed
branches for soil, wastes, down spouts or ·trap ventilation
shall be of wrought iron and connected, as shown, to sewer.
All connections through which water is to pass shall be
made with Ys, and these Ys or elbows used to change from
vertical or horizontal runs or to receive water closets, shall
be cut to the proper grade of horizontal pipes.

Branches to water closets and slop sinks shall be run in
floor, finishing at floor level with a flange proper to receive
each fixture set level and true.

Safe waste risers shall be located as shown, with lower end
closed with safe waste valves of brass, loosely hung.

All risers shall be clamped securely to beams or walls at
every second floor.

Branches for ventilation of traps shall be run at such a
height as to give a continuous rise to the connection from
crown of trap. Where a branch crosses a room or corridor
on any floor, it shall be run to the floor above and dropped
to fixture. In all cases, whether of waste or ventilating
pipes, the iron shall go through the floor or plaster, leaving the opening at finished floor or plaster line ready for
connection.

HAND HOLES.—Hand holes of same diameter as
pipes shall be put in on every riser at bottom, and each

branch to sink shall have one or more hand holes. Hand holes shall also be put in wherever indicated. All these openings shall be closed with brass trap screws as specified.

TRAPS.—Each fixture shall have a separate trap, as hereinafter specified, set in all cases as close to outlet as practicable.

WATER SUPPLY.

The system of supply shall furnish both hot and cold water to all basins and sinks, and cold water only to all urinals and water closets, and cold water for supplying drinking fountains.

The surge tank, hot water tank, house tank, drinking water tank and pumps will be furnished and set as elsewhere specified, and provided with flanged openings, and the work included in these specifications will commence at this point.

STREET CONNECTION.—The contractor shall connect to the city water main and lay in a three inch cast iron water pipe and run same to meter. All work of opening street, replacing street, pipe laying, valves and joints shall be done in compliance with the Rules of the Street and Water Departments of Fort Wayne, and subject to their approval.

METERS.—The contractor shall furnish, set and connect where shown or directed a three inch Worthington Water Meter, or other equally as good, approved by the Commissioners and Architect, and shall connect the same to water supply, and from meter shall run a galvanized iron pipe on the basement ceiling as shown. The contractor shall set a double faced gate valve on each side of meter. Also a Pratt & Cady check valve on inlet to meter.

SURGE TANK CONNECTIONS.—The contractor shall connect the three inch supply pipe to surge tank with a three inch straightway gate valve for shutting off tank. Also set and connect on surge tank a three inch cross head with similar gate valve for shutting off tank, and from the cross head, take off the main supply pipes as shown, running around with galvanized iron pipe on basement ceiling to supply the risers. All of these pipes shall be provided at header with a brass gate valve and an L. H. finished shower stop dripping into a copper trough connected to catch basin, and with a brass tag attached to each stop, plainly marking the destination of supply. In addition to the connections shown, there shall be left on cross head one three inch by one inch tee plugged. The contractor shall also connect the drain of surge tank with the catch basin with valve on same, and shall connect the header on surge tank with supply from house tank with straightway valve on same.

HOT WATER TANKS.—The contractor shall connect the hot water tanks to City pressure supply with straightway valve on connection to tank, and cross connect City pressure supply to hot water tanks to supply from the house tank with valves on same, all substantially as shown on the plan. Also set on tank a two inch cross head (with one extra plugged one inch opening) with similar gate valve on connection to tanks, and from header take off the supply connections as shown on plan. All these supplies shall be provided at header with valves, shower cocks, copper trough with drip pipe and brass tags, all as specified for surge tank connections. The tanks shall also be connected with one and one-half inch drain pipe with stop valve, to catch basin or sewer. All return hot water mains shall be returned and connected into boiler as shown with Pratt & Cady check valve on return preventing back flow of cold water.

HOUSE TANK CONNECTIONS.—The contractor shall connect the house pump to house tank with straightway valve and Pratt & Cady check valve on connection to pump. This pipe to continue up in tank seven feet; also take off supply from house tank with straightway valve on same and run to basement as shown. Also connect the three inch overflow from top of tank to nearest down spout. Also connect the two inch drain from bottom of tank into same overflow pipe, with valve connection to tank. The pipe from pump to tank shall be cross connected to City pressure supply pipe and suitably valved so that the tank may be supplied direct from the City pressure, if the same is strong enough. Also run a one inch Tell-tale pipe from house tank down to and over sink in basement.

PUMP CONNECTIONS.—The circulating pump shall be connected to cooling tank and to drinking water supply pipe with straightway valve on connection to each. The house pump shall be connected to surge tank and to house tank with straightway valve on connection to each.

The circulating pump and house pump shall be cross connected and suitably valved with straightway valves, all substantially as shown on plan.

FILTER CONNECTIONS.—The contractor shall connect the filter to City pressure supply and to cooling tank with straightway valves on connection to each. The supply to filter to be cross connected to the supply from house tank and suitably valved so that either may be used.

RISERS.—All risers shall be galvanized wrought iron pipes securely clamped to walls or beams at every second floor. Each riser shall be provided with wheel handled shut-off valve at bottom. Each hot water riser shall be made with offsets or other satisfactory manner so that expansion at fixture shall not exceed one-eighth inch. At the bottom of each descending return hot water riser shall be set a swinging check valve to prevent back flow.

BRANCHES.—Branches shall be of wrought iron carried through the floor or outside plaster line. All branches run below the floor line or in partitions are to be safed.

PIPE COVERING.—All hot water supply mains and returns in basement, and hot water supplies in attic, are to be closely covered with an approved covering equal to "Bradley" insulated air covering, thoroughly secured.

CONNECTIONS.—All connections to fixtures shall be of lead or N. P. brass, as specified with fixtures. Connections shall be as follows:

To wash basins, one-half inch pipe.

To closet cisterns, one-half inch pipe.

To urinal cisterns, one-half inch pipe.

To sinks, three-quarter inch pipe.

To drinking fountains, one-half inch pipe.

SUPPLIES AND WASTES IN GENERAL.—All branches or horizontal pipes shall be laid to drain. All branches from soil or waste pipes shall rise one-quarter inch to one foot in length unless otherwise directed by the Architect. All risers shall be set so as to be conveniently incased. The hot water supply mains in basement shall have a uniform rise from header at hot water tank to farthest riser. The hot water returns shall have a corresponding fall from farthest riser to hot water tank.

FLASHING.—All pipes passing through roof shall be flashed at roof line with a piece of sheet lead not less than eight inches in all directions from pipe, and a piece of lead soil pipe one-eighth inch thick running from sheet lead to top of pipe and turned over inside pipe one inch. The sheet lead and pipe shall be joined with a wiped joint.

SAFING.—All horizontal wastes and supplies in floors, above the ground room floor, shall be safed, and all horizontal supplies run in partitions and all supplies run in attic shall be safed. The contractor shall line the gutters with sheet lead neatly formed and turned over the edges of the gutters and closely nailed with copper nails. The gutters shall be of sufficient size to carry away the water in case of burst, and shall be connected to safe wastes with one and one-quarter inch lead pipe with funnel shaped connection to gutter. The gutter shall be fitted with covers of No. twenty-six galvanized iron, soldered at the joints and turned down at the edges one inch at each side. The contractor shall also suspend a tin or copper gutter directly underneath the drinking water pipe that crosses in the pump room in basement, and shall connect this gutter with three-quarter inch pipe to catch basin.

The contractor shall provide and properly put in place a sheet lead pan for collecting all drip from the house tank as installed in the attic The pan to be at least seven feet in diameter, with edges turned up six inches; all joints to be perfectly water tight; the whole to be well supported to prevent any sagging; and to be drained by a one and one-quarter inch wrought iron pipe carried to the nearest downspout and properly connected thereto.

STREET WASHERS.—Where shown, the contractor shall set and connect three-quarter inch street washers with iron box case and loose key, all to be approved by the Architect. The iron case shall be set flush with side walk level.

DOWN SPOUTS.—The contractor shall furnish and set the down spouts of wrought iron pipe where shown as specified for wrought iron risers and shall connect same to sewer. Each down spout shall be provided with an approved deep trap at bottom; down spouts shall be connected to gutter with heavy copper wire.

HOSE CONNECTIONS.—Where shown, the contractor shall connect three-quarter inch brass sill cocks with loose keys for hose connection. Cocks to be brought just through wall about one foot above floor; also set sill cocks in basement where shown.

TESTING.—All supply pipes shall be stopped off at

outlets and filled with cold water at normal pressure. Sewers shall be stopped off at building wall and all openings on sewers, wastes, soil and ventilation pipes shall be filled with water to roof. All gutters shall be stopped off and filled with water. All the foregoing shall be carefully examined and if any are found to leak, they shall be made tight and the test repeated. These tests shall be made in the presence of the Architect, and the contractor shall not be entitled to demand or receive payment, nor shall the final certificate be issued until all this work be made as tight as contemplated herein.

FIXTURES.—All the fixtures included in this specification are to be set without casing, and the contractor shall set the same and the connections thereto in a neat, finished and uniform manner.

WATER CLOSETS.—Where shown, the contractor shall set and connect Huber's plain vitreous china "Sylvan" Pneumatic water closet, Plate 411, Huber's Cat. with design G, hardwood cistern, twenty-four inches by twelve inches by seventeen inches in depth, or other equally as good closet and hardwood cistern, approved by Commissioners and Architect. Closets shall be set on specified flange, joint packed with a rubber ring and closet bolted to flange, with brass bolts, with nickel-plated heads. Closets shall be furnished with Appendo seats and covers attached to bowl, except closets in marble stalls shall have seats only. Cisterns shall be set on nickel-plated brass knee brackets on board and cistern connected to bowl with nickel-plated brass air pipe, and one and one-half nickel-plated brass flush pipe with combination nickel-plated brass pipe and seat stop plate. Cisterns shall be connected to supply with top nickel-plated one-half inch brass supply to wall with wall plate, and with lever handled stop on same.

The closet in toilet room off of Commissioners' Court Room, Office floor, is to have a one-half inch nickel-plated supply pipe from floor with nickel-plated brass floor plate and wheel handled stop cock. Each closet is to be vented with a two inch vent connection to vent riser.

URINALS.—Where shown, the contractor shall set and connect in marble stalls, No. One Bedfordshire lip urinals, furnished with Yeteve hardwood cabinet finished, automatic flushing tanks, No. seven hundred and four, Walcott & Hurlbut's Catalogue, or other equally as good urinals and automatic syphon tank approved by Architect. Urinals shall be ground to fit back and held in place with brass bolts with nickel-plated heads. Each urinal shall be fitted with nickel plate brass urinal trap, connected straight through back to concealed waste. Each group of three urinals shall have a three gallon Yeteve tank as before described with nickel-plate brass tripple connection and divided nickel-plate brass three-quarter inch flush pipe with rubber expansion coupling to each urinal. The cistern shall be supported on nickel-plate brass brackets. Each cistern shall be supplied with a one-half inch nickel-plated brass supply pipe, with nickel-plate lever handled stop on same.

SINKS.—In basement, where directed, the contractor shall set and connect one square enameled iron sink, eighteen inches by thirty-six inches, set on bronzed iron brackets and connected to sewer with two inch lead trap and pipe, and sink fitted with two three-quarter inch brass flanged bibbs, (one hose bibb) connected to cold and hot water. Elsewhere where shown, throughout the building, the contractor shall set and connect enameled iron slop sinks with trap Standards, eighteen inches by twenty-two inches by twelve inches with brass strainer, polished cast brass rim, enameled iron back with air chambers and nickel-plated "Fuller" faucets as shown in "D" eight hundred and eighty-four, Wolff's Catalogue, or other equally as good, approved by the Architect. All exposed supplies and vent connections to be nickel-plated brass. The sink in mens' private lavatory (office floor) to be finished zinc white finish outside.

WASH BASINS.—In the toilet rooms of Court Rooms No. One and Four, (Court Room floor) and in Judge's and Witness room, opposite Court Room No. Two, where shown, the contractor shall set and connect one and one-quarter inch countersunk Italian marble lavatories, thirty-two inches by twenty-two inches, with fancy shell soap cup and sixteen inch high back, fourteen inches by seventeen inches oval basins, and nickel-plated brass slab brackets, "D" five hundred and thirty-eight Wolff's Catalogue, or other equally as good, approved by Architect. Basins shall be fitted with nickel-plated brass all over ideal wastes, nickel-plated brass standard trap, with pipe to floor, and vent to wall nickel-plated brass supplies to floor with improved air chambers and key stops, and "Cornell's" nickel-plated self-closing basin cocks, or other equally as good, approved by Architect.

Where shown in ladies' toilet rooms and mens' public toilet room, (Ground floor) and mens' private toilet room, (Office floor) the contractor shall set and connect fourteen by seventeen inch oval bowls in Italian marble slabs, twenty-two inches wide, four feet, six inches long, one and one-half inch thick, with sixteen inch backs and ends, and five inch apron on front and free end. Basins to be fitted with nickel-plated self-closing basin cocks, nickel-plated all over ideal wastes, nickel-plated brass traps, supply pipes, air chambers and stops, all as before described. The waste of traps and the supplies to go to the wall or floor as the case may require. The free end of slab shall be fitted with nickel-plated brass leg, similar to that shown in plate five hundred and one Wolff's Catalogue, and apron shall have nickel-plated brass apron supports at wall.

The three hole slab in the mens' private toilet room, (Ground floor) shall have similar bowls and be trimmed as before described, but the slabs shall be six feet, nine inches long, and shall have an additional nickel-plated brass leg at center. All of these slabs are to have fancy shell soap cups cut in them. Elsewhere throughout the building the contractor shall set and connect similar basins, slabs to be twenty-two inches by thirty inches, with sixteen inch backs and ends, and with nickel-plated brass bracket on free end, as before described; the free end of slab to be rounded similar to that specified for toilet rooms for Court Rooms. All other trimmings to be as before described, the waste and supplies to go to wall or floor as the case may require.

MIRRORS.—The contractor shall furnish and set beveled plate glass mirrors, set in three inches wide marble frames, over lavatories in mens' and womens' toilet rooms, (Ground floor); mens' toilet room, (Office floor); over lavatories in Judges' private toilet rooms, and in attorneys' room near Law Library.

The mirror frames are to be thirty inches high, and of the same length as the marble slabs. Contractor to make a neat joint between marble mirror frame and marble back.

DRINKING FOUNTAINS.—Where shown, the contractor shall set and connect Italian marble drinking fountains, twenty inch by fourteen inch by eighteen inch back, similar to "D" two hundred and fifty-eight Wolff's Catalogue. The slab is to be countersunk one-half inch and furnished with nickel-plated drip cup, with shield nickel-plated brass brackets; nickel-plate brass urinal trap as shown in "D" nine hundred and ninety-seven Wolff's catalogue, and special nickel-plated self-closing basin faucet, as shown in "D," two hundred and fifty-six Wolff's catalogue. Other quality as good drinking fountains and fittings approved by Architect may be substituted for the above.

MARBLE.—The contractor shall set and furnish the marble backs, slabs and aprons necessary for all wash basins, and the marble stalls and marble base for urinals, and the marble stalls for water closets and marble for drinking fountains. The marble shall be selected white Italian clear ringing when struck, highly polished on all exposed edges and surfaces. The marble, unless otherwise specified, shall be cut strictly in accordance with detailed drawings, which will be furnished as required, with fine joints, and shall be set in plaster where it goes against walls, and in all cases, shall be secured with concealed fastenings, and set plumb and true. The marble for basin slabs shall be of the dimensions already specified. The backs, ends and aprons shall be seven-eighths inches thick, and of the dimensions before specified. All edges of slabs shall be beveled, and the backs shall be properly stopped off. Marble stalls for urinals shall be five feet eight inches high, partition shall be two feet center to center, and two feet in width; the middle partitions to be fitted with twelve inch nickel-plated brass leg; the end partitions to go to floor; partitions and backs to be seven-eighths inches thick, marble bases to be two feet two inches wide, and of the length required, and two inches thick, countersunk one-half inch. The marble for water closet stalls shall be as shown in detail drawings, except that backs in womens' private and public lavatories, (Ground floor) shall be of the width shown. Marble slabs for drinking fountains shall be one and one-quarter inches thick, and of the dimensions before specified; the backs and sides shall be seven-eighths inches thick.

The marble work herein specified shall be considered with that portion of the general specifications for closet and lavatory wainscoting, and this work shall conform to it.

BRASS WORK.—The contractor shall furnish and set all brass work for marble water closet stalls and urinal stalls, included herein. It shall be first quality, and shall be heavily nickel-plated. He shall set the clamps for fastening the door strips to marble fronts of water closet stalls; clamps to be furnished as specified.

POWER STATION.

In addition to the work already shown and specified, the contractor shall connect to the sewer of Jail and lay in a six inch vitrified pipe sewer to a point two feet outside of Power Station, and shall there connect with a six inch extra heavy soil pipe and run the same into Power Station to the location of soil and waste risers and to the catch basin.

All this work to be done in the same manner as before described.

The contractor shall furnish and set in front of boilers where directed an iron catch basin similar to the one before described, and shall connect this basin to sewer. Also furnish and set in boiler room of Power Station a two inch water meter with valved connections to and from meter, all similar to that before specified. Also connect to City water main and lay in a two inch water pipe, and connect same to meter, and from meter take off branches to supply the various plumbing fixtures, and provide two three-quarter inch hose connections, one in front of boilers and one near coal room where directed. The contractor shall run all necessary supply pipes, vent pipes and soil pipes and make all necessary connections for plumbing fixtures of engineer and firemen, and he shall also connect the down spouts of roof to sewer, all in a manner as above specified.

FIXTURES.—In toilet rooms the contractor shall furnish and set closets, same as already described, the closets to have hardwood seat and cover, and fittings as previously specified. Also set and connect Italian marble lavatories twenty inches by thirty inches with sixteen inch back and end; slabs fitted with fourteen inch by seventeen inch bowls and connected to wastes with one and one-quarter inch nickel-plated traps and nickel-plated vent connection. Bowls shall be fitted with two self-closing basin cocks connected to cold and hot water with nickel-plated brass supplies with air chambers.

In fireman's room where directed, the contractor shall

set and connect one long oval, flushing rim hopper closet, furnished with seat and cover, hardwood tank, tank connected to closet with one and one-quarter inch lead flushing pipe, with clips and protector, and cistern connected to cold water supply with one-half inch lead pipe. Also set and connect one thirty-six inch porcelain lined iron sink, same as before described, connected to waste with two inch lead trap and pipe, and provided with two three-quarter inch brass sink bibbs, (one hose bibb.)

GUARANTEE.—The contractor shall be responsible for all plumbing, and shall make good and repair at his own expense, as may be necessary, all defective work, material or fixtures, which may show itself within one year from the date of final certificate, for which time he shall guarantee all plumbing work to be in good order and repair, unless broken or injured by improper use.

HOT WATER SUPPLY TANKS.

The contractor to furnish and install two three hundred gallon tanks, each to be fitted with a steam coil; the dimensions of coil shall be such as to raise the temperature of three hundred gallons of water per hour from a temperature of forty degrees Fahr. to at least one hundred and eighty-five degrees Fahr. Steam to be furnished from the steam heating main at atmospheric pressure. Each tank to be provided with a two-inch flanged inlet and a two-inch flanged outlet opening, these to be located at opposite ends of the tank at the bottom and top respectively; the plumbing contractor making all connections to these openings. The steam coils to be connected to heating main by the steam heating contractor; at the bottom of the tank at the opposite end to the circulating intake the contractor in each case will have a one and one-half inch opening for the piping contractor to attach his drain pipes.

Each heating tank to be provided with a thermometer attached so as to indicate the true temperature of the water in the tank. Tanks each to be provided with suitable manhole, with plate, yoke, bolts and gasket complete. Tanks to be approximately three feet diameter by six feet long. Tank heads to be dished.

The tanks are to be set in cast iron saddles furnished by the contractor, one saddle to be placed near each end of the tank; saddles to be securely fastened to brick piers provided as specified; the contractor providing all anchor bolts, plates and nuts complete; tanks to be so installed that the end from which the drip connections are taken will be at least one inch lower than the opposite end. Care must be taken to provide a proper draining of the steam coil.

The contractor will thoroughly cover each of the tanks with a two-inch sectional covering of either "Magnesia," "Asbestos Sponge" or "Nonpareil" cork.

HOUSE TANK.

The contractor to furnish and erect in the attic in a position substantially as shown on the drawings and make complete in all details, ready for pipe connections, one fifteen hundred gallon tank, the tank to be set upon supports provided by the contractor, so as to elevate the tank to the highest possible degree. Tank to be approximately six feet in diameter by eight feet high; to be built of three inch selected cypress, and to be fitted with a suitable number of bands provided with patent tightening lugs; foundations for supports to be as elsewhere specified. The contractor to properly level the tank; a tight fitting wooden cover will be provided to permit of access to tank; the tank shall be thoroughly tight throughout. The contractor to make provision for all piping connections, same to be made with proper flanges: A two inch drain at bottom, a three inch overflow within ten inches of the top, a two inch for house supply at bottom, a one and one-quarter delivery from pump. This latter to be placed in the center of the bottom of the tank.

Contractor to thoroughly paint tank with three coats of weather-proof paint as approved by the Architect.

ELECTRIC PUMP.

The contractor shall furnish and install in the basement of the building, in a position substantially as shown on the drawings, one electrically operated house pump having a capacity of delivering at least two hundred and fifty gallons of water per hour against a minimum head of seventy-one feet; the pump to be the "Worthington" or other make equally good and acceptable to the Commissioners and Architect.

The motor to be securely bolted to the cast iron base of pump, the whole to be rigidly fastened to a foundation as elsewhere specified, and built in accordance with templates to be furnished by the contractor immediately after the signing of the contract.

In his proposal the bidder shall specify the style and make of pump he proposes to furnish; the type and make of motor; whether motor is belted, geared or directly connected to plunger shaft, and give horse-power capacity of the motor.

REGULATING DEVICE.—The contractor shall furnish with his motor one "Whittingham" solenoid automatic motor starter of the latest and most approved design; connections shall be so made as to cause the motor to be brought into operation when the level of the water in the tank has been lowered not more than eighteen inches, and to automatically cut the motor entirely out of the circuit when the tank is filled.

Arrangements shall also be made so that the motor may operate independently of the regulating device or be controlled entirely by such device, as the case may be; the float installed in the tank to be connected so as to operate a double pole snap switch or other device as may be approved by the Commissioners and Architect.

The contractor to furnish all labor and material for the proper installing of his motor, as called for, and to make all electrical connections complete from the distributing board in the basement to his motor, and between the motor and the switch at the tank.

GUARANTEES.—The contractor shall guarantee in connection with the "General Requirements" all his apparatus as proposed to satisfactorily perform the work for which it is installed, and to deliver two hundred and fifty gallons of water per hour under a maximum head of seventy-one feet.

That the combination shall operate noiselessly, without pounding or vibration when performing its full duty.

That the entire equipment shall be of the best workmanship, material and finish throughout, and to replace and make good any defects which may develop in his apparatus under proper and normal working conditions within one year from date of final completion of his contract, without cost to the County.

SURGE TANK.

The contractor shall furnish and install in the basement

of the building, in a location substantially as shown on the drawings, providing suitable cast iron saddles or supports all as may be required, one fifteen hundred gallon closed surge tank; tank to be of one-quarter inch best tank steel, to be single riveted throughout with dished heads; tank to be horizontal and approximately five feet in diameter by ten feet long.

The contractor to provide all openings suitably flanged to make piping connections thereto, as follows: Three inch intake in one head; three inch out-take at top of tank at the opposite end to the intake; two inch drip to be located in the bottom of tank at the opposite end to the intake; a two inch outlet in bottom of tank at same end as intake; this for the house pump suction.

REFRIGERATING APPARATUS.

The contractor will furnish and install a complete and comprehensive system of refrigeration for the cooling of drinking water. The system to include a "Jewell" filter; cooling tank complete with pipe coils and flanged outlets ready for plumbing attachments to circulating pipe; ammonia compressor complete with all attachments; accessories, regulating devices and fittings with all connections to pipe coils in cooling tank; circulating pump with connections made to cooling tank; and electric motors complete as called for herein.

The refrigerating machine to have a capacity of cooling one hundred and twenty-five gallons of water per hour continuously for ten hours from seventy degrees to forty degrees Fahr. Compressor to be placed upon foundations as specified and built in accordance with templates furnished by the contractor, the contractor providing all anchor bolts, plates and nuts complete.

The cooling tank to have a capacity of four hundred gallons; to be vertical and approximately three feet eight inches in diameter by five feet in height; to be built of best one-quarter inch tank steel single riveted throughout. Tank to be entirely covered by a two-inch sectional covering of either "Magnesia," "Asbestos Sponge" or "Nonpareil" cork.

The filter to be that known as the "Jewell" filter, and to be of sufficient size to thoroughly filter at least one hundred and twenty-five gallons of water per hour, the contractor providing flanged inlet for plumbing attachments. The contractor to make all connections from filter to cooling tank.

With his apparatus the contractor will provide one electrically operated circulating pump to be of the "Worthington" or other make equally good and acceptable to the Commissioners and Architect, securely fastened to foundation as specified and built in accordance with templates provided by the contractor. The pump to be capable of pumping five hundred gallons of water per hour; the motor to be of ample capacity to perform the work intended; to be of the "Holtzed-Cabot," "Crocker-Wheeler" or "Lundell" make; motor to be securely bolted to the cast iron bed plate to which the pump is attached. The preference will be given to a belted or rotary pumping outfit.

With the electric pump contractor will provide and install at some point convenient to the motor one approved non-combustible starting rheostat and double-pole, marble base, polished brass or copper knife switch; making all connections to motor complete and ready for attaching mains to the switch.

The contractor will also furnish and install one electric motor for driving the ammonia compressor, foundation to be provided as specified and built in accordance with templates to be furnished by the contractor. Motor to be of sufficient capacity to perform the work for which it is intended without over-loading. With this motor there will be provided a sliding base and adjustable belt tightener, one approved incombustible starting rheostat and double-pole, marble base, brass or copper knife switch; contractor to properly install the above, making all connections to motor complete and ready for the mains to be attached to the switch. This motor to be mechanically and electrically perfect; the armature to be perfectly balanced; and to run noiselessly and without vibration. Motor to be provided with self-oiling bearings.

Contractor will furnish one endless, light-double leather belt for driving the ammonia compressor; belt to be made of positive center cuts, strictly short lap, and to be perfectly balanced throughout.

The contractor will furnish and properly install the above apparatus, together with all accessories going to make it a perfect system complete and ready for operation, locating his apparatus substantially as shown on the drawings.

It is intended that these specifications cover an entire system for accomplishing the purposes as herein specified, and the bidder will make his proposition accordingly, as it is understood that after the completion of the contract a successful system will be expected and demanded of the contractor.

The contractor guarantees his complete refrigerating system as installed by him, to perform the work for which it is installed and to cool at least one hundred and twenty-five gallons of water per hour from seventy degrees to forty degrees Fahrenheit, continuously for ten hours. That the ammonia compressor will operate noiselessly and without vibration. He further guarantees that none but the best workmanship and material will be used throughout, and to replace and make good any part of his entire system as installed, which may develop defects within two years from date of final acceptance of his work, without cost to the County.

ELECTRICAL CONSTRUCTION.

The intent of the following specifications is to provide for the complete Electrical Construction for light and power, also furnishing of and wiring for telephones, all as hereinafter set forth and specified.

GENERAL.—Wiring for light and power shall comprise everything from and including the two main fuse blocks on the Distributing Board in basement, (which Board will be considered the center of distribution,) to the outlets inclusive, including cut-out cabinet with cut-outs in place, as hereafter specified, also properly connecting and mounting all snap switches, but it does not include the furnishing of switches or fixtures, except switches on cut-out board in basement, switches in vault in Treasurer's office and pendants with buttons and key sockets in attic and basement, as hereinafter set forth.

A detailed drawing showing size of Distributing Board and arrangement of cut-outs, switches, etc., must be submitted by the contractor, to the Architect and be approved by him before work on same is begun. The exact location of Board will be indicated by Architect.

The contractor shall also submit to the Architect for his approval a drawing showing dimensions of cut-out cabinets.

ARRANGEMENT OF LIGHTING SYSTEM.—Wiring to be done initially for two thousand seven hundred and seventy-three sixteen candle power lamps; lamps all to be operated from direct current generators of proper E. M. F., to give one hundred and ten volts at the lamps.

The building has been divided into eight main sections and a separate riser or risers provided for each section, from which riser all lamps in corresponding section shall be fed.

The sections are plainly marked upon the drawings and clearly show the number of lights taken from the respective cut-out cabinets on each floor.

All wiring throughout the building to be what is known as "concealed work," using an approved iron conduit, with the exception that in the basement, the feeders to risers may be run on porcelain or glass knobs to cut-out cabinets in basement at foot of risers.

The entire construction is to be done in comformity with the letter and spirit of these specifications and the latest Rules and Regulations of the National Board of Fire Underwriters.

WIRING.—In all wiring throughout the building there shall be no joints in any Feeder or Main excepting where the gauge of same changes; in case of top circuits, the wire shall be continuous from cut-out to outlet, joints being permitted only at outlets and in junction boxes, wherever it be necessary to install.

FEEDERS AND MAINS FOR LIGHTING.—The general arrangement of feeders and risers will be as follows:

The contractor will furnish and place a slate Distributing Board, located in basement near point where riser No. Two starts, as shown on plans, and said Distributing Board will be considered the center of distribution for the building.

This Board to be made of Vermont slate of high insulating quality and enameled. Slate to be not less than one and one-quarter inch thick. The board to be of ample size to prevent crowding of such fuse holders or switches as may be specified hereafter.

From this Board feeders will be run to the base of the various risers as hereinafter described. From these points, risers will run from and through one cut-out cabinet to another as located on each floor. The feeders and mains in basement are to be carried on porcelain knobs on ceiling, and are to be rigidly supported and held in place without the use of tie wires.

If any feeder or main should be in danger of coming in

contact with any iron beams, pipes or ventilating ducts, it must be carried around in approved conduit and be so arranged as to be free from danger of coming in contact with same.

The Distributing Board is to be properly mounted on standards of light T irons, which irons shall run from floor to the ceiling of basement and shall be securely fastened at both points.

The arrangement on the Board will be as follows:

The two feeders from station are to run to two main fuse blocks, thence to a double pole knife blade switch (of not less than two thousand five hundred amperes carrying capacity, thence to two copper bus bars of two and one-half square inches in cross section.

. From these bus bars will be run feeders to the various riser points as hereinafter stated.

Also four separate circuits, one for the Dome Lighting, one for Statue Lighting, one for Clock Lighting and one for False Dome Lighting.

In addition to these circuits there will also be run from the Distributing Board the various power circuits, as hereinafter described, under the heading of "Wiring for Power."

Each pole for each circuit leading from the Distributing Board is to be protected with a suitable fuse holder, provided with removable screw plug mounted on Distributing Board.

All connections to bus bars to be made by solid bus bar connections; the busses being fastened to the Board by means of standard bus bar stands.

Conduits enclosing all risers will be carried in recesses provided in the walls of partitions, as shown on the drawings.

FEEDERS.—Feeders in the basement to be run from and to the following points, and each leg will be of the size specified:

From Distributing Board a feeder to Riser No. 1 of 500,-000 circular mils; continued to foot of Riser No. 5 by 4-0 B. & S. Gauge.

From Distributing Board a feeder to Riser No. 2 of 500,000 circular mils.

From Distributing Board a feeder to Riser No. 3 of 1,000,000 circular mils. Continued to foot of Riser No. 7 by 4-0 B. & S. Gauge.

From Distributing Board a feeder to foot of Riser No. 4 of 1,000,000 circular mils.

Continued to foot of Riser No. 8 by 4-0 B. & S. Gauge.

From Distributing Board a feeder to foot of Riser No. 6 by 4-0 B. & S. Gauge.

RISERS.—All risers to run from cabinets in basement to cabinets on Court Room floor.

Each leg of the risers to be of the following sizes:

Risers Nos. 1, 2, 3. and 4, 500,000 circular mils.

Risers Nos. 5, 6, 7, and 8, 4-0 B. & S. Gauge.

The sizes of the feeders to Dome, Clock Tower, etc., will be as follows:

A circuit of No. 2 B. & S. Gauge wire to Dome for inside gallery lighting of Dome.

A circuit of No. 2 B. & S. Gauge wire to lights for False Dome.

A circuit of No. o B. & S. Guage wire to clock tower for supplying of lights for clock dials.

A circuit of No. 2 B. & S. Guage wire to upper part of Dome to supply lights for Illuminating Statue.

The contractor is to specially note that all risers are to be CONTINUOUS from the basement cut-out cabinet to the cut-out cabinets on Court Room Floor.

Risers will, in all cases, connect with a main double pole branch cut-out in each respective cabinet.

From each cut-out cabinet on Court Room Floor on risers Nos. 1, 2, 3 and 4 three circuits will be run, as before stated, to a convenient point for switch box to be located in the adjacent Court Room, and from there to the attic, where suitable cut-out cabinets will be placed, from which will be distributed all the lights in and over said Court Room, with the exception of the two five-light clusters in each alcove, the two seven-light chandeliers and the one light outlets on the Judge's bench. These lights on the Judge's bench will be fed from the main cut-out cabinet in Corridor and be carried in Court Room Floor up through Judge's bench for a standard. The lights in alcoves of Court Rooms No. 1 and 4 may be taken off risers Nos. 5 and 8, with switch outlets near doors at end of east corridor. The lights in alcove of Court Rooms Nos. 2 and 3 to be on separate switch circuit from the other lights in said rooms; the switch oulet to be in the switch cabinet with the other three switches for said rooms respectively.

The three switches which are to be placed in each Court Room will control all the ceiling lights in and over its respective room, one switch to control all lights over the skylight; one to all lights in Court Room, bordering or encircling the skylight, the other switch the balance of ceiling lights, excepting those in connecting toilet room, alcoves or lights in small rooms adjoining the Court Room as above specified.

Each of the two seven-light clusters on Judge's bench to be provided with a switch outlet, said outlet to be on the Judge's bench or at some convenient point to be designated by the Architect.

Sizes of mains which are to be run to cut-out cabinet in attic over their respective Court Rooms, from which the Court Room lights are to be supplied as previously described:

From riser No. 1 to cut-out cabinet over Court Room Floor, a No. 6 B. &. S. gauge main for lamps over ceiling light.

From riser No. 1 to cut-out cabinet over Court Room Floor, a No. 2 B. & S. gauge main for 92 lamps around ceiling light.

From riser No. 1 to cut-out cabinet over Court Room Floor, a No. 2 B. &. S. gauge main for balance of ceiling lights.

From riser No. 2 to cut-out cabinet over Court Room Floor a No. 6 B. &. S. gauge main for lamps over ceiling light.

For riser No. 2 to cut-out cabinet over Court Room Floor a No. 2 B. &. S. gauge main for 86 lamps around ceiling light.

From riser No. 2 to cut-out cabinet over Court Room Floor a No. 2 B. &. S. gauge main for balance of ceiling lights.

From riser No. 3 to cut-out cabinet over Court Room Floor a No. 6 B. &. S. gauge main for lamps over ceiling light.

From riser No. 3 to cut-out cabinet over Court Room Floor a No. 2 B. &. S. gauge main for 64 lamps around ceiling light.

From riser No. 3 to cut-out cabinet over Court Room Floor a No. 2 B. &. S. gauge main for balance of ceiling lights.

From riser No. 4 to cut-out cabinet over Court Room Floor a No. 6 B. &. S. gauge main for lamps over ceiling light.

From riser No. 4 to cut-out cabinet over Court Room Floor a No. 2 B. &. S. gauge main for 90 lamps around ceiling light.

From riser No. 4 to cut-out cabinet over Court Room Floor a No. 2 B. &. S. gauge main for balance of ceiling lights.

PUBLIC LIGHT SWITCHES.—All switches for public lights in Corridors, Vestibules, Rotunda, etc., are to be placed in the cut-out cabinet from which these lights are to be fed. but if, especially in the case of cabinets on risers No. 2 and No. 6, there be not ample room to place the necessary switches in these cabinets, in addition to the cut-outs and use the same sized cabinet as elsewhere in the building, then the contractor shall furnish and place a switch cabinet in close proximity to the cut-out cabinet and all the switches intended for said cut-out cabinet shall be placed in this switch cabinet. All switches in cabinets to be approved double pole snap switches.

The drawings will indicate which specific outlets are controlled by the various switches other than those lights in Court Rooms.

BRANCH CIRCUITS.—The wire for branch or tap circuits shall be of such size that the maximum loss between any cut-out cabinet and any lamp fed from that cabinet will not exceed two volts with all lamps on the circuit burning; this loss to be measured between the main cut-out and the lamp socket.

There shall not be over 10 lamps of 16 candle power, or their equivalent in incandescent lamps of other candle power on any branch circuit, except by special written permission of the Architect. The exact location of all lamp outlets and switch outlets to be indicated by the Architect before work is begun. All lamp outlets, however, will be located approximately as shown on the drawings; the switch outlets for the various offices will usually be located near entrance of office.

CUT-OUTS.—Each cut-out cabinet will contain the branch cut-outs for all the lamps to be fed from its respective riser.

There shall be a double branch main cut-out located between vertical riser and the branch cut-outs.

All main cut-outs to be porcelain or slate strip cut-outs with all wire connections made on the face of the cut-outs.

All branch cut-outs to be plug cut-outs.

All cut-outs must be properly fused with fuses of the proper carrying capacity to protect the circuit in which it is placed.

All main line fuses to be copper tipped.

At each riser point in basement, there will be a cut-out cabinet, in which will be placed the necessary main and branch cut-outs.

For such lighting as is necessary in the basement, the contractor will use such of these various points for distributing centers as he may find most convenient.

SNAP SWITCHES.—All switch outlets, unless otherwise specified, are to be for flush snap switches.

Contractor to make list of and properly tag with a small brass tag, all switch outlets, marking same with number of lights to be controlled from same to prevent any mistakes when switches are attached.

All switch outlets for more than five 16 c. p. lamps, or their equivalent are to be double pole.

SWITCHES FOR VAULT IN TREASURER'S OFFICE.—There are to be two one light outlights in the vault; one light in upper and one in lower vault. Each of these lights are to be controlled by a separate switch; these switches to be so arranged that the light in its respective vault will be turned on and off automatically by the opening and closing of the vault doors.

CUT OUT CABINETS.—On all floors, recesses will be left for the cut-out cabinets.

The cut-out cabinets to be furnished and installed by the contractor; cabinets to be of slate, the backs and walls to be not less than three-quarters of an inch thick.

The contractor will furnish the doors, hinges and locks for the cabinets, which must be set with front of same flush with the plaster line.

The holes through which the conduit passes in entering the cabinet must be of such size as will make a neat, snug fit, and the tubing be rigidly fastened.

The cabinets will be of sufficient depth to allow at least a two inch space between any cut-out or switch and the inside of door when closed.

The cabinets are to be of ample size to prevent any crowding of cut-outs, and preferably, of uniform size throughout on Ground, Office and Court Room Floors, being approximately 24x24 inches outside measurements.

The contractor before placing the cabinets will confer with the Architect or his representatives as to the exact location of same.

SPECIAL CIRCUITS TO DOME.—At the ends of the mains to Dome, over false Dome, to Statue and to Clock Tower, a suitable cut-out cabinet of slate is to be installed for each, with main and such branch cut-outs in same as each case may require.

On the Distributing Board in basement there will be furnished and placed by the contractor four double pole knife switches of suitable carrying capacity, one switch to control the inside Dome lights, one the lighting over false Dome, one the Statue lights and one the Clock lights.

WIRING IN GENERAL.—All wires to be of soft drawn copper of at least 98 per cent. conductivity.

All wires used throughout shall be of the highest grade of what is known as "rubber covered wire."

The tap circuits for such lighting as may be called for for general lighting in the atttic, may be taken from the most convenient cut-out cabinet on Court Room Floor.

The cabinets provided over dome and in attic for Court Room lighting will be of slate.

The contractor is to provide flexible lamp pendants installed complete with Edison or T. H. base key sockets and "peanut" cord adjusters for all basement and attic lighting, except Jury Rooms in attic.

A "P. & S." porcelain ceiling button shall be provided for supporting each lamp.

As before stated, while the lamp oulets will be located essentially as shown on drawings, and the switch outlets as stated in these specifications, it is understood that the Architect shall have the right to determine the exact location of any or all outlets and that any changes in position of outlets that the Architect may order before the work has been installed shall not be the basis of an extra charge.

The wires at each switch and lamp outlet are to extend at least nine inches beyond the wall.

In the running of conduits the runs must be continuous from outlet to outlet, and the joints carefully, securely and tightly made. No wires to be drawn in until after all conduit has been installed, except by WRITTEN permission of the Architect.

The contractor will find a space of not less than two inches between the floor tile and the heavy eye beams, which space he may use for the carrying of conduits.

The contractor is to provide all outlets with suitable support for brackets, switches or fixtures, as the case may require.

All circuits may be run with two wires in the same conduit wherever the wire is not larger than a No. 10 B. & S. guage. More than one wire larger than a No. 10 B. &. S. gauge must not be run in the same conduit unless special permission is received from the Architect.

In the case of all floor outlets, as for desks or counter lights, in the Commissioner's Court Rooms, Surveyor's, Recorder's. Treasurer's Offices, etc., the iron conduit is to extend at least one foot above the finished floor and is to be rigidly fastened in place. The wires at such outlets to . extend at least three feet from end of conduit.

TAPPING CIRCUITS TO RISERS OTHER THAN SPECIFIED.—In case there are certain distributing or tap circuits which the contractor desires to connect with a different riser than specified, he shall do so only upon obtaining written permission in each specific case from the Architect.

KNIFE SWITCHES.—All knife switches are to be polished brass or copper, and to be so designed that the current is not carried through the joints; blades and contact pieces to be of pure rolled copper and to have a contact surface of at least one square inch for each fifty amperes.

CONNECTIONS.—All connections, including portions of switches, fuse holders, etc., carrying current, shall be of such material and such dimensions that their conductivity shall be at least equal to that of an equal length of the wire or cable of the circuit into which they are inserted.

ELEVATOR ELECTRIC LIGHT CABLES.—The contractor shall furnish and install three electric light elevator cables: the cable to be that as manufactured by the Western Electric Company, catalogue No. 2097, or other make of cable equally good in the opinion of the Architect.

A separate circuit shall be run for each elevator from the most convenient cut-out cabinet of Office Floor to an oulet provided in the elevator shaft. To this outlet the contractor shall properly attach the elevator cable and also properly connect cable to outlet on elevator car.

GROUND FLOOR.

	Light Outlets.								Bracket Outlets.	Switch Outlets.
	1	2	3	4	5	6	7	8-lt.	1 2-lt.	
Sheriff's Office				4			1		1	1-12-lt. 1- 7-lt.
Cell	1									
Private Elevator Passage	1								1	
Clerks' Storage	6									2- 6-lt.
Janitors' & Engr's Room			4							1-12-lt.
Clock Room				1					2	
Janitors' Hall Way									2	
West Vestibule	16	24	2							1- 6-lt. 1-24-lt. 2-16-lt. 1- 8-lt.
County Assessors'				2						1- 8-lt.
Auditor's Storage		8								2- 8-lt.
Township Assessors'		1		4			1		1	1- 7-lt. 1-16-lt.
Township Trustee'				4			1			1-12-lt. 1-16-lt. 1- 7-lt.
Assembly Room				15					1	3-20-lt.
Women's Lavatory	1	1								1- 3-lt.
East Vestibule		8	11			1				1- 9-lt. 1-28-lt. 1-21-lt.
Women's Private Lavatory		3		2						1- 6-lt. 1- 3-lt.
Janitors' Room		1								
Men's Private Lavatory	4	3								1- 6-lt. 1- 4-lt.
Men's Public Lavatory	5									1- 5-lt.
Coroner's Room				4					1	1-16-lt.
County Superintendent's				4			1		1	1- 7-lt. 1-16-lt.
No. Vestibule Sec. 1		12	6							2- 6-lt. 1-30-lt.
North Lobby Sec. 2		3	14							1- 6-lt. 1-42-lt.
Rotunda		20	8							2-12-lt. 4-10-lt.
So. Lobby Sec. 3		3	14							1- 6-lt. 1-42-lt.
So. Vestibule No. 4		12	6							2- 6-lt. 1-30-lt.

OFFICE FLOOR.

	Light Outlets.								Bracket Outlets.		Switch Outlets.
	1	2	3	4	5	6	7	8-1t.	1	2-1t.	
Surveyor's Office....(floor).. ..											
		2	4	..	2						1-10-1t.
Clerk's Document Room............		4									1- 8-1t.
Clerks' Work Room..				5					1		1- 5-1t.
											1-20-1t.
Passage		1							1		
Clerk's Office ....(floor) 8 ..		8							1		
			3								2-12-1t.
Bridge Superintendent's..........				2							1- 8-1t.
Commissioners' Passage	2										
Private Room........		1									
Commissioner's Court Room ....(floor).. ..											
		22			2				2	2-1t.	1-12-1t.
											1-32-1t.
Commissioners' Lavatory		1							1		1- 2-1t.
Commissioners' and Auditor's Rooms. ..				4			1				1- 7-1t.
Auditor's Room Sec. 3 ..(floor).. ..											1-12-1t.
			4	8							2-12-1t.
Auditor's Room Sec. 4 ..					4		1				1- 7-1t.
											1-20-1t.
Auditor's Work Room 1	1				12						3-20-1t.
Treasurer's Office Sec 4 and 8 ....(Ceil.).. ..											
	6	1	2		5						1- 5-1t.
..(floor).. ..											1-20-1t.
		1									2- 1-1t.
Treasurer's Office Sec 7..................(floor).. ..											1- 6-1t.
		3		10							2-16-1t.
East Lobby..........		8	2		2						1-14-1t.
											1-18-1t.
Men's Private Lavatory 8	8										2- 4-1t.
Recorder's Office Sec. 6. ....(floor).. ..											
		2		8					1		2-12-1t.
Recorder's Office Sec. 5........		4			5						1- 5-1t.
											1- 8-1t.
											1-20-1t.
North Lobby..........		12	8								2- 2-1t.
											2-16-1t.
											1-30-1t.
Rotunda.............. 40	40		4								2- 6-1t.
											4-10-1t.

OFFICE FLOOR.—CONTINUED.

	Light Outlets.								Bracket Outlets.	Switch Outlets.
	1	2	3	4	5	6	7	8-lt.	1 2-lt.	
Main Hallway by Commissioner's Court Room...... ..		..	2	..						2- 3-lt.
South Lobby......... ..	12	8	..							1- 2-lt.
										1-16-lt.
										1-30-lt.

COURT ROOM FLOOR.

	Light Outlets.								Bracket Outlets.	Switch Outlets.
	1	2	3	4	5	6	7	8	15 lt. 1 lt.	
Court Room No. 1...........	92	1	..	..	..	16	2	..		2 7-lt. 3 switches for Ceiling.
Court Room Alcove.........	..	..	..	..	2	..	..	..		1 10-lt.
Jury Room.................	..	..	..	2	..	..	..	..		1 8-lt.
Ante Room................	1	..	..	..	..	..	..			
Toilet Room..............	..	..	..	..	..	..	..		1	
Judge's Chamber......	..	4	..	..	1	..	..	..	1	1 5-lt.
										1 8-lt.
Alcove....................	..	1	..	..	..	..	..			1 2-lt.
Hallway..................	..	..	..	..	..	..	..		2	
Judicial, Public and Communicating Corridors.....	..	10	24	..	..	5	..	..	Newel Post. 2	3 10-lt. 2 15-lt. 1 12-lt. 2 9-lt. 2 21-lt. 1 20-lt.
Sheriff's Office..............	..	..	..	1	..	..	..			1 4-lt.
Detained Witness Office	..	..	..	1	..	..	..		1	1 4-lt.
Judge's Chamber...........	..	4	..	..	1	..	..		1	1 5 lt.
										1 8-lt.
Law Library and Alcoves, including Gallery.........	..	4	14	..	..	4	..	..	1	2 10-lt. 4 6-lt. 2 12-lt.
Jury Room and Passage in Attic	2	..	..	..	4	..	..	..	2	2 10-lt.
Judge's Chamber...........	..	4	..	..	1	..	..		1	1 5-lt.
										1 8-lt.
Stenographer's Room.......	..	..	1	..	..	..	..		1	1 4-lt.
Witness Room..............	..	..	..	1	..	..	..		1	1 5-lt.
Judge's Chamber in Alcove.	..	5	..	..	1	..	..		1	1 2-lt.
										1 5-lt.
										1 8-lt.
Jury Room.................	..	..	..	2	..	..	..			1 8-lt.
Ante Room................	1									
Toilet Room..............									1	
Court Room No. 4...........	90	1	10	..	16	..	2	..		2 7-lt. 3 switches for Ceiling.

COURT ROOM FLOOR.—Continued.

	Light Outlets.								Bracket Outlets. 15 lt.	Switch Outlets. 1 lt.
	1	2	3	4	5	6	7	8		
Court Room Alcove					2					1 10-lt.
Consultation Room				2						1 6-lt.
Witness Room					1				1	1 4-lt.
Witness Room					1				1	1 4-lt.
Prosecuting Attorney				2						1 6-lt.
Grand Jury Room					3				1	1 10-lt.
										1 5-lt.
Janitor's Hall										2
Ante Room				1						
Attorney's Room				1						1 4-lt.
Witness Room				2						1 6-lt.
Consultation Room				2						1 6-lt.
Witness Room				1					1	1 3-lt.
Witness Room					1				1	1 4-lt.
Consultation					2					1 6-lt.
Court Room No. 2	86	1				6	2	6	2	7-lt.
										1 10-lt.
										3 Switches for Ceiling.
Witness Alcove					2					1 10-lt.
Ante and Toilet Room										2
Consulation Room			1							
Rotunda							8			1 16-lt.
										1 40-lt.
Court Room No. 3	64	1			20		2		2	7-lt.
										3 Switches for Ceiling.
Alcove					2					1 0-lt.
Consulation Room			1							
Ante and Toilet Room										2

ATTIC.—Not including Jury Rooms or Balcony of Library.

16-1 lt. outlets distributed.

2-5 lt. outlets distributed.

2-2 lt. outlets distributed.

2-5 lt. switch outlets to be located on Court Room Floor.

AROUND SPRING OF DOME.—96-1 lt. outlets.

BASEMENT.—50-1 lts. distributed.

In addition to the above, there will be 15-1 light outlets for lighting over ceiling lights, over upper staircase hall. Same to be controlled by one switch located on the Court Room Floor at cut-out cabinet at riser No. 6.

Also 25-1 light outlets over ceiling light in Library, same to be controlled by one switch located in cut-out cabinet at riser No. 2.

EXPLANATION OF DRAWINGS, ETC.—There are eight lighting drawings.

The Ground Floor plans in each case, with the exception of the basement showing the number of outlets, the number of lamps to outlets, the number of switches and the specific lamps to be controlled by each switch, with the exception of switches for Court Rooms, as previously stated.

The ceiling plans of Ground, Office and Court Room Floors are simply to indicate more accurately the exact location of ceiling outlets; no brackets or floor outlets being shown on same.

Contractor will note that in the separate circuit run for lights over False Dome, to Clock and to upper part of Tower for lighting of Statue, he is simply to run circuit to cut-out box located at some convenient point to where lamps are to be supplied and is to equip this box with the necessssary cut-outs, but he is to run no tap circuits from these points. This will be done later by the Fixture contractor.

The contractor will also note that the four circuits leading from Court Rooms to supply lights over the ceiling lights, are also to be simply carried to cut-out box and supplied with the necessary cut-outs by another contractor, namely the Fixture contractor, who will run circuit from these points.

While ceiling outlets are indicated on the drawings by two characters. one a cross and the other a circle, contractor will note that whenever a ceiling outlet is indicated by a cross it indicates that a chandelier will be hung at that point; where indicated with a circle, that the lamps will be placed close to the ceiling, no chandelier being used.

The location of all lights in the basement to be determined later, but the contractor is to install 50-1 light outlets complete with drop cord, socket and "peanut" adjustor, he taking these lights from the most convenient cut-out boxes at the foot of any riser.

While the specifications state that the building shall be wired initially for 2773 lights, the contractor is to base his figures as to the number of lights and outlets from the tabulation given and from the drawings.

WIRING FOR POWER.—The wiring for the power circuits shall be as follows:

The contractor shall furnish and install complete all circuits, switches and cut-outs, as hereinafter specified, with the exception of such switches as may be specifically stated are to be furnished and installed by another contractor.

From the Distributing Board there will be run two circuits or feeders to cut-out cabinets in attic; said cabinets to be placed in attic near riser points 2 and 3. Each of the circuits will be of No. 2-0 B. &. S. gauge cable.

From the cut-out cabinet near riser point 2, there shall be run four circuits of No. 6 B. &. S. gauge wire, connecting respectively with the three motors in North end of attic and the motor in West Center.

From the cut-out cabinet near riser No. 3 there shall be run three circuits of No. 6 B. &. S. gauge wire, connecting with the three motors in the South End of attic.

Each of these circuits to be properly fused and properly connected to the motor and such appliances as may be furnished with or attached to the motor.

From the Distributing Board is to be run to each of the Elevator motors on East side of building a circuit of No. o B. & S. Guage wire.

From the Distributing Board a circuit of No. 6 B. & S. guage wire is to be run to the Elevator motor, which is to operate the Private Elevator.

From the Distributing Board a circuit of No. 4 B. &. S. gauge wire is to be run to each of the motors driving duplex fans in basement.

Each of these circuits is to be properly connected to bus-bar on Distributing Board in the same manner as the feeders for Lighting Circuits, excepting such circuits as hereinafter described.

The contractor is to furnish two twenty ampere double pole approved snap switches and attach same to the distributing board and properly connect these switches to switches provided and installed as elsewhere specified, one near electric pump house; the other near cold water circulating electric pump.

The contractor shall also furnish and install on the Distributing board one fifty ampere double pole knife switch, connecting same with the electric motor driving Ammonia Compressor.

The contractor shall put in place and securely attach to the Distributing Board the three double pole "Cutter" Automatic Magnetic Crcuit Breakers furnished under elevator specifications.

The contractor will also install nine "Whittingham" Solenoid Automatic Motor Starters, as specified; these motor starters to be placed near their respective motors in some convenient location; two of them to be installed in the basement for controlling the motors operating duplex fans; the remaining seven to be installed in the attic, in connection with the motors driving exhaust fans as there located.

The contractor will make all connections from automatic motor starters to their respective motors and from starters to double pole snap switches.

The switches in connection with the starters for motors in attic to be placed in the most convenient cut-out cabinets on the Court Room Floor. The switches in connection with the motors to duplexed fans to be placed on Distributing Board.

The contractor will furnish and install the necessary double pole porcelain cut-outs.

The same general specifications governing kind of construction, grade of material, etc., ior lighting, shall apply to and be considered as specifications governing construction of power circuits.

The sizes of all wires and the carrying capacity of all switches and bus-bars have been based on operating all light and electric power on a direct current at 110 volt pressure.

In case a different system of voltage be selected, as indicated in the Dynamo Specifications, the specific sizes of most all mains and feeders and the carrying capacity of the switches will be re-proportioned to correspond to the change in voltage or system and the necessary changes be also made in arrangement of the Distributing Board, etc.

In such case the sizes of all feeders and mains and the carrying capacity of switches and arrangement of Distributing Board shall be submitted for the approval of the Architect and Commissioners and be made satisfactory to them and receive their written approval before any work is begun.

TELEPHONE CONSTRUCTION.—It is the intention of these specifications to govern the furnishing and installing of a complete house telephone syetem, as hereinafter outlined, ior facilitating communication between offices in different parts of the Court House, Power Station and Sheriff's residence.

There will be twenty-five instruments installed. Four of these will be desk sets; these latter instruments to be furnished for Judge's chambers.

These instruments will comprise transmitter, receiver and magneto. and will be so designed that any instrument may be put in circuit with any other instrument, as is specifically stated hereaiter.

The finish of all instruments will be white oak.

The telephones which this contractor proposes to furnish will be such as to meet with the approval of the Commissioners and Architect.

For telephone wire the contractor will use double and twisted telephone rubber-covered conductor under single braid. All wires will be concealed, and wherever running behind furring to be fastened by staples, using a saddle under same.

Wherever wires are run in the cement floors or on any walls where there is no furring, they will be drawn into "plain interior conduit." No restriction will be made as to size of tubes to be used, with the exception, however, that no tube will be larger than one and one-half inches in diameter when laid in the floor, while those on walls will necessarily be of such size as to permit of their being sufficiently covered by the plastering, and if necessary two or more tubes shall be used. The number of wires that may be drawn into one tube will be such that there will be no crowding, and that any one wire may readily be withdrawn should occasion require.

The exact location of all telephones will be decided upon by the Architect. The right to change the location of any or all instruments is reserved by the Architect, but any such change must be made before the work has been installed, and under such circumstances shall not be the basis of an extra charge.

The contractor shall guarantee the complete telephone system throughout to work successfully; that none but the best workmanship and material will enter into his work, and further, to replace and make good any defects in workmanship or material which may develop within ninety days from the final acceptance of his work, providing such defects develop under proper and normal conditions of operation, without cost to the County.

The location of all telephones, with a list of those to which each instrument is to be capable of being connected, is given herewith.

Court Room Floor:
One in each Court Room.
One in each Judge's Chamber.
One in Prosecuting Attorney's office.
One in Library.

These ten instruments to communicate with the instruments to be installed in the Clerk's office on the Court Room floor · with the Sheriff's and the Janitor's instruments located in their respective offices on the Ground floor; and also with the instruments located in the Clerk's office on the Office floor.

OFFICE FLOOR.—Each office on the Office floor to have an instrument; all instruments to communicate with one another and to have connection with the instruments in the Jantor's office on the Ground floor. The Janitor's instrument will therefore connect with all instruments on the Court Room floor and with all instruments on the Office floor.

GROUND FLOOR.—The instrument installed in the Sheriff's office will be connected to the instrument installed in the Sheriff's residence. The Janitor's instrument will be connected to an instrument to be located in the Power Station. By this arrangement the Sheriff's instrument connects with all instruments on the Court Room floor and the instrument in the Sheriff's residence, while the Janitor's instrument, in addition to being connected with telephones as already described, will also connect with the instrument in the Power Station.

WIRING FOR SHERIFF'S HOUSE AND JAIL.— From switchboard in Power Station the contractor shall run two five hundred thousand circular mil lead covered cables, (of the same kind as used in tunnel) to some suitable distributing point in Sheriff's house. The approximate distance between switch board and this point is one hundred and fifty feet; this cable to be carried in suitable iron or wooden ducts; contractor to properly connect this cable to switch on switch board, said switch being provided as specified.

At the end of feeder in Sheriff's house contractor will place a suitable distributing cut-out cabinet made of slate, which cabinet shall be led the necessary mains and tap circuits. The Sheriff's house and Jail will be wired for not less than one hundred 16 c. p. lamps and the necessary switch outlets.

All wiring in Sheriff's house shall be done in oak moulding. All wires of Jail shall be carried in iron armored conduit; said conduit to be securely fastened to wall or ceiling as the case may require.

The contractor shall state the amount for which he will run the cable, furnish and mount cut-out cabinet and the necessary cut-outs, as above stated, and shall also state the amount he will charge per 1 light, 2 light, 3 light, 4 light and 5 light outlet with moulding work; the same per switch outlet with the necessary snap switches furnished and in place; and shall also state the amount he will charge per 1 light 2 light, 3 light, 4 light and 5 light outlet, running same in iron armored conduit, the same per switch outlet; the wires to switches being carried in iron armored conduit.

The Jail is attached to Sheriff's house, the dimensions of Sheriff's house being approximately 40x40 feet, (two stories). The Jail being approximately 90x40 feet.

For all wiring a high grade of rubber covered wire shall be used, and the "General Conditions" of the specifications covering the Electrical construction in the Court House shall govern the work above described.

WIRING FOR POWER PLANT.—All wires in Power Station shall be carried in iron armored conduit. For all ceiling lights conduit will be carried over the ceiling. For all bracket lights, conduit is to be concealed in walls.

The General conditions of the specifications covering the Electrical Construction in the Court House, as to the grade of material, character of work, etc., shall govern the work in Power Station.

LOCATION AND DISTRIBUTION OF LIGHTS.— In the boiler room, at west end of boilers. there shall be 2-1 light pendants; on either side of boilers 1-1 light pendant. The contractor to furnish and install the necessary drop cord for these outlets.

At the east end of boilers, there shall be 2-3 light ceiling outlets. These outlets to be controlled by a switch located at some convenient point in boiler room.

ENGINE ROOM.—There shall be 6-2 light bracket outlets, 2 on north, 2 on south and 2 on west walls.

There shall be 4-10 light ceiling outlets, each outlet to be controlled by a flush snap switch located near door leading into boiler room.

Over Visitors' Gallery there shall be 2-4 light outlets, These lights to be controlled by a switch located in Engineer's room.

At entrance leading to Visitors' Gallery there shall be 1-5 light outlet, controlled by a switch located in Engineer's room.

In the office there shall be 1-2 light ceiling outlet controlled by a switch located in the office, and 1-1 light bracket.

In Engineer's room there shall be 1-2 light ceiling outlet controlled by switch and also 1-1 light bracket.

All switch outlets to be arranged for flush switches.

In addition to the above the contractor shall wire for 6-1 light wall outlets located in boiler room at such points as may be determined by the Architect, and furnish and install at each of these outlets, an Edison Keyless receptacle, polished with a D. P. fuse, same as listed on page 117 of their supply catalogue, No. 50,979. Each of these receptacles to be fused for 1 light.

LIGHTS FOR GAUGES ON BOILERS.—There shall be 6-1 light outlets for supplying light in front of each of the three steam gauges and each of the three water gauges. In front of these gauges the contractor shall furnish and install a key socket and reflector half shade. The circuits to be suitably run in on iron conduit and each light rigidly supported with a Crowfoot; lamp so placed as to properly light its respective gauge.

MAIN CUT-OUT CABINET.—From the switch board, under the floor, there shall be run a No. 6 circuit to the cut-out cabinet, which cut-out cabinet will be located either in the Engineer's room or at a point near door leading into boiler room. This cabinet to be made of slate. The contractor to furnish and install the cabinet complete with the exception of the doors and hinges for same.

The contractor is to furnish all labor and material as above specified for the wiring of the Station complete, but is not to furnish figures or switches unless specifically mentioned.

ELECTRIC ELEVATORS.

GENERAL—The intent of the following specifications is to provide for the furnishing and installing of the required Elevators, as hereinafter set forth, complete and ready for service.

The bidder is requested to make his proposition on the entire equipment as outlined herein, and in accordance with the General Requirements.

NUMBER, TYPE AND CAPACITY.—The contractor shall furnish, erect complete in all details and deliver in perfect working condition, three electrically operated, worm geared Passenger Elevators and all necessary appliances for a complete apparatus in every particular.

Two of these elevators will be used exclusively for public passenger service and the remaining one for private service.

Exclusive of the weight of the car, the Passenger Elevators to each have a carrying capacity of two thousand two hundred pounds at a speed of one hundred and fifty feet per minute.

The private elevator to have a carrying capacity of one thousand two hundred pounds at a speed of one hundred and fifty feet per minute.

DIMENSIONS AND TRAVEL.—The dimensions of the Passenger Elevator hatchways are five feet four inches by seven feet five inches

Those for Private Elevator, four feet by six feet.

The platforms of the different elevators are to be made to suit the hatchways provided, but dimensions shall, in case of Passenger Elevators, be five feet by six feet, and for the Private Elevator as near as possible to be three feet nine inches by four feet six inches.

The distance from the top of each floor to the top of the next floor above is given below:

Ground floor to Office floor, fifteen feet two inches.

Office floor to Court Room floor, seventeen feet six inches.

The ceiling of Court Room floor is twelve feet six inches above the floor line.

The total travel of each car will be from the level of the –

Ground floor to the level of the Court Room floor, a distance of thirty-two feet eight inches.

The contractor must compare these dimensions with those on the drawings, and notify the Architect of any discrepencies.

GUIDE POSTS AND GUIDE STRIPS.—The Passenger Elevators to be provided with side guide posts. Private Elevator with corner guide posts.

The contractor to furnish and install all guide posts and guide strips. Guide posts to be made of steel tees, weighing about sixteen pounds per foot and planed so the stem of the tee will form a true and uniform guide; tees to be securely bolted to the iron structure of the building.

SHEAVES AND BEAMS.—The contractor is to provide suitable main overhead sheaves and furnish and install suitable steel "I" beam supports for same; "I" beams being supported upon castings furnished by the contractor; these castings to be securely bolted to the "I" beams and to the iron structure of the building.

The contractor will find upon referring to the drawings, that ample head room for elevator sheaves and supports is allowed in the details of the building construction.

CABLES.—Each of the Passenger Elevators will be operated by six five eights inch diameter, patent flattened strand rope, as manufactured by A. Leschen & Sons Rope company, St. Louis, Mo., or other strand rope equally good and acceptable to the Commissioners and Architect. Cables to be made of Swedish iron with hemp center; the counterbalances for each elevator to be attached to two of these cables.

For Private Elevator, four cables, three-quarter inch in diameter, will be furnished; cables to be the same as those furnished for Passenger Elevators; two of these cables to be used for counterbalances.

COUNTERBALANCES.—Each of the elevators furnished by the contractor shall be properly counterbalanced by independent counterweights, so as to obtain the best possible results; weights shall run in guides made of "T" iron, same to be furnished and substantially bolted in place by the contractor.

CONTROLLING AND SAFETY DEVICES.—Each elevator shall be operated by a wheel or lever device of such design as to insure the safe and easy controlling of the car.

Elevators will be furnished with the latest and most perfect, modern automatic safety appliances for the protection of the system proposed to the satisfaction of the Architect.

Elevator doors to have anti-friction rollers six inches diameter.

PASSENGER CARS.—The contractor will furnish and properly attach to platform of each Passenger Elevator, one wrought iron Passenger Car, of such design and finish as to meet with the approval of the Architect and equal to what can be bought by the expenditure of $450.00. The back and sides of cars to be enclosed; the wainscoating to be old bronze finish and to be approximately three feet six inches high with a six inch grill work border on top; above this border the finish to be plain with the exception of the scroll work for the corners of pannels, which is to correspond with the other work, and to extend at least two feet six inches above the top of the border; the back of car to have a beveled plate mirror approximately four feet long by two feet six inches high, to be securely fastened in place with a suitably designed border all around; the front of the car to be entirely of grill work with the exception of the three feet opening for doorway, the panel to be made in accordance with designs to be furnished by the Architect. The corner posts of the car to be made of wrought iron pipe not less than two inches in diameter. The entire top finish of the car to be open grill work of such design as to meet the Architect's approval, the design to be submitted by the contractor before purchasing same.

For the Private Elevator contractor will furnish one wrought iron Passenger Car of neat design and finish equal to what can be bought by the expenditure of $250, to the satisfaction of the Architect, the contractor submitting cuts showing the general design of car to be furnished, same for the approval of the Achitect

FOUNDATIONS.—The contractor will, in each case, provide suitable and substantial concrete foundations for the reception of his apparatus; foundations to be made of Portland cement concrete; the entire motor equipment to be secured firmly by proper anchor bolts and plates, all to be furnished complete and in position by the contractor.

ANNUNCIATOR.—The contractor will provide with each elevator which he furnishes, one annunciator with vibrating bell and three indications; the annunciator to have black dial, nickel plated pointers and white letters, reading as follows

"Ground Floor," "Office Floor," "Court Room Floor."

The finish of annunciator case to be white oak. The contractor to place annunciator in the car as may be directed by the Architect, and shall furnish and install cable, batteries and push buttons complete ready for operation. The push buttons furnished to be plain polished, dark bronze, screw cap; the approximate diameter of button being two inches.

MOTOR EQUIPMENT.—Each motor, as furnished, will be provided with self oiling bearings of approved design; armatures to be perfectly balanced; machines to run smoothly and quietly and without vibration under any and all changes of load up to their full rated capacity.

With each motor equipment there will be furnished one double pole "Cutter" automatic magnetic circuit breaker of suitable carrying capacity; the finish to be "Standard No. 1" (switchboard type direct current).

AIR CUSHION.—For safety there must be air cushions at the base of each Passenger Elevator shaft and the contractor will furnish detailed drawings, which will clearly show just what he proposes for such work. Such drawings must be approved by the Architect before commencing the work.

CONCLUSION.—The entire system of elevators, which the contractor furnishes, to be installed in the most approved manner, to be complete in all details and ready for operation.

The platform of all cages to be of first quality seasoned oak, and iron construction; the flooring to be seven-eighth inch selected hard maple. The cage and car to be securely fastened to the platform.

The contractor shall use all due diligence in hastening the completion of his work, in the time stipulated in his contract, and with this end in view, he will familiarize himself with other work that may in any way be dependent upon or connected with his own.

INSPECTION.—The contractor, after the completion of his work, shall have each separate elevator equipment inspected by some reputable insurance company and he shall also furnish the Commissioners with a certificate of inspection and insurance policy for each elevator, for one year, same to be fully paid by the contractor.

GUARANTEES.—The contractor guarantees by the acceptance of these specifications that each of the elevator equipments he is to furnish will not consume more than 9,000 watts of electrical energy when starting with full load in the car, either going up or coming down; or when carrying full load and under full speed, either going up or coming down; or more than 8,500 watts of electrical energy when empty, with the exception of the operator, and under full speed, either going up or coming down.

The contractor further guarantees that each motor will run without injurious sparking at the commutator throughout all changes and conditions of load.

That the motor and apparatus may run continuously for twenty-four hours without undue heating or endangering the apparatus in any way.

That every part going to make up the complete elevator system shall be erected in a substantial manner; the best workmanship, material and finish throughout, and to operate smoothly and without hitch or jerk throughout all ranges of travel, and also the perfect operation of all elevators

The contractor to hold himself responsible for any defects which may develop in any part of his system as installed, due to faulty wormanship or material of any sort whatsoever, and to replace and make good any such faulty parts which may have developed defects within one year from the date of final acceptance of his work, without cost to the County.

FINALLY.—It is intended that these specifications shall cover the complete and entire equipment for three separate elevators, for accomplishing the purpose as herein specified, and for the purpose of furthering his own interests, the bidder is requested to carefully examine the drawings and acquaint himself with all conditions, as that after the completion of the contract, a complete and perfect apparatus in every particular will be expected and demanded of the contractor.

ELEVATOR DOOR OPENING MECHANISM.— The intent of the following is to provide for the furnishing and installing complete in all particulars, making all attachments to elevator doors of a device for automatically opening and closing elevator doors.

The contractor will furnish the hydraulic door opening mechanism as manufactured by the Burdett-Rowntree Manufacturing Company, or other hydraulic mechanism equally good and acceptable to the Commissioners and Architect.

There will be furnished with each of the Passenger Elevators a mechanism to be operated by hydraulic motor, the complete mechanism to be so designed as to automatically open elevator doors located at each floor; there being nine doors in all, three for each elevator. The operation of the mechanism shall be such that the opening of the doors will be entirely under the control of the operator in the car; while the closing of these doors shall be automatic, positive and entirely beyond the control of the operator, the door closing immediately upon the car starting either up or down, and not to open before the car has reached the floor level. The levers operating to open and close the door to be so arranged that when the door is closed it will be impossible for them to be opened from the outside of the elevator shaft.

The hydraulic motor will be located at some convenient point in the basement, at or near its elevator shaft; the motor to be so designed as to operate under the minimum pressure of 20 pounds per square inch; contractor doing all piping and making all connections between the motor, and the plumbing pipe leading from the house tank, and from motor to city water main; placing brass bodied straightway valves in each pipe as well as directed in the supply pipe to the motor; the connections so arranged as to take the supply from either source.

The entire equipment throughout to be positive in its action; the doors to open and close noiselessly and to operate smoothly, without hitch or jar, such operation being absolutely essential on account of the close proximinty of the Elevators to the Court Rooms.

Contractor shall install the complete apparatus in every particular for accomplishing the work as outlined herein, and shall guarantee that only the best workmanship and material will be used throughout, and to replace and make good all parts going to make the completed system, which may develop defects due to faulty workmanship or material within one year from the date of final completion of his work, without cost to the County.

CLOCK SYSTEM.

The intent of the following specifications is to provide for a complete and comprehensive Clock System, and contemplates the furnishing and installing complete in all details of the Master Clock, Secondary Clocks, Tower Clock, compressors, tanks, valves, gauges and all piping and fittings necessarily involved, all as hereinafter set forth and specified.

All bids submitted shall cover the entire system complete and in operation. All apparatus or appliances essential to the proper and convenient operation of the apparatus or system proposed shall be furnished and installed without extra charge, notwithstanding each and every item necessarily involved in the work is not specifically mentioned.

SYSTEM.---The system shall be the "Johnson Pneumatic Time System," or other equally good and acceptable to the Commissioners and Architect.

MASTER CLOCK.—In the clock room on the ground floor there shall be placed a Master Clock, which shall govern all the Secondary Clocks distributed throughout the building, and also the Tower Clock. The Master Clock is to be enclosed in a glass and iron case; the case to be so designed that the works and pendulum can be clearly seen. The center of the dial to be not less than five feet six inches from the floor or base of the clock. The glass sides, etc., of the case to be the finest French plate with beveled edges. The iron skeleton frame to be of plain but neat design.

ACCURACY.—The Master Clock shall have an accuracy of within six seconds per month, without being synchronized.

DIAL.—The dial of the Master Clock to be fourteen inches in diameter and made of the finest French plate glass, unground, excepting a ground ring some two and one-half inches broad, which shall be behind the figures, the figures to be in gold.

SECONDARY CLOCKS.—There shall be thirty-two Secondary Clocks in the rooms as hereinafter stated, this number not including the Tower Clock. Each of these clocks to be so set that the dial is nearly flush with the finished wall, and to be enclosed in a neat metal frame, round in form and of such finish as may be approved by the Architect.

DIALS.—The dials of all Secondary Clocks to be of the

finest French plate glass, with beveled edges, the finish of same to be determined by the Architect. The diameter of all Secondary dials to be fourteen inches, excepting in the Court Rooms, where they will be of the diameter indicated on drawings.

VARIATION.—The variation between the Master Clock and any Secondary Clock shall not exceed fifteen seconds.

COMPRESSORS AND FITTINGS.—There are to be furnished and installed two hydraulic air compressors with suitable pressure tanks and all necessary valves, tanks, etc., said compressors to be placed in a convenient place in the basement of the Court House; either compressor to be of ample capacity to supply waste air on the time system.

Each compressor is to be furnished with a suitable pump governor, which will automatically stop the compressor when the desired limit of air pressure is reached and start it again when the pressure has fallen the desired amount. The starting pressure on one compressor to be about one pound less than on the other, so that in case the high pressure or regular compressor should fail to work, the one set for lower pressure, or the reserve compressor, will be automatically cut in and do the work. Water from the house tank with a minimum pressure of twenty pounds will be used to operate the reserve compressor; the other compressor to be operated direct from City water pressure, which pressure will not be less than twenty-five pounds.

AIR PIPING.—All the air piping except in basement to be concealed and to be of galvanized iron pipe excepting where the construction demands armored lead tubing.

The contractor to make all the necessary attachments to water pipe in the building on one side and on the waste side, carry pipe to catch basin, located as shown on drawings.

IN GENERAL.—The exact location of the clocks will be decided upon by the Architect, but the plans show approximately the location of each clock.

The clocks will be located as follows:

GROUND FLOOR.—One clock in Township Assessor's office.

· One clock in County Assesssor's office.

One clock in Sheriff's office.

One clock in County Superintendent's office.

One clock in Assembly Room.

One clock in Township Trustee's office.

OFFICE FLOOR.—One clock in Auditor's office.
One clock in Commissioner's Court Room.
One clock over Commissioner's door in Rotunda.
One clock in Clerk's office.
One clock in Recorder's office.
One clock in Treasurer's office.

COURT ROOM FLOOR.—Four clocks, one in each of the Judge's chambers.
One clock in Sheriff's and Clerk's office.
One clock in Grand Jury Room.
One clock in Prosecuting Attorney's Room.
Four clocks, one in each Court Room.
Four clocks, one in each jury room.
Four clocks, one at each end of corridors.
One clock in Library.
Total, 32.

TOWER CLOCK.

DIALS.—In the tower there will be placed a Secondary Clock to operate each of the four dials, of diameter shown on drawings. Dials to be etched glass, not less than one-half inch in thickness, and mounted in a suitable iron skeleton frame. A sketch showing the design of dial shall be submitted to the Architect for approval.

HANDS.—Hands are to be made of sheet copper, hollow and of elliptical form in cross section; the frame of hands being an iron truss frame; the center of hands to be of bronze and to have hexagonal tapering bores, which shall fit accurately over tapering bronze spindles; the hands are to be accurately counterbalanced.

GENERAL.—In case a single clock movement is used to operate all four dials, said movement is to be housed in a neat and accessible case by the contractor, and placed in the center of Clock Room in tower. Cluster movements to be placed on each dial and same to be operated through bevel gear and metal shafting from center movement.

GUARANTEE.—All piping is to be guaranteed to be practically air tight under twenty pounds pressure.

The contractor shall guarantee that the material and workmanship for this entire system is to be of the best quality throughout, and that accurate and reliable service will be obtained under normal usage, and that he will replace and make good any defects in workmanship or material which may develop within one year from the date of final acceptance of his work, without cost to the County.

BOILERS.

NUMBER AND TYPE.—The contractor to furnish and erect complete, ready for operation, three water tube boilers.

ARRANGEMENT.—These three boilers are to be set in one battery; the stack being located as near as possible directly behind the boilers, substantially as shown on the drawings.

HEATING SURFACE AND EVAPORATION.— Each boiler shall have about thirteen hundred square feet of heating surface, and shall be capable of evaporating thirty-nine hundred pounds of water per hour from one hundred degrees, into steam of seventy pounds pressure, when operated under normal working conditions, and shall be capable of being forced fifty per cent. over the above rated capacity without injury.

Natural gas is to be burned under all of these boilers and the furnaces must be designed as to give the capacity required above, under such conditions.

TRIMMINGS AND FITTINGS.—All front and cleaning door castings will be neatly and tightly fitted.

Each boiler shall be provided with the latest and most modern improvements and appliances; all necessary mountings and trimmings complete, including "Reliance" water column, with high and low water alarm; all requisite valves, gauges and piping. The "Reliance" water column to be complete with one three-quarter inch finished self-closing water guage and three three-quarter inch "Lunkenheimer" self-grinding, rotating gauge cocks provided with chain to facilitate opening and closing.

The safety valves to be furnished with these boilers to be three and one-half inch nickel seat, "Consolidated" patent lock-up pop safety valves. The steam gauges to be "Crosby" manufacture, with brass cases, eight inch dial, black face, white figures; gauges to register from one to one hundred and fifty pounds, with proper sub-divisions.

With each boiler there shall also be furnished one five inch "extra heavy" Chapman angle stop valve for the main steam outlet; one copper funnel for the gauge cocks with drip connections to ash pit; one two and one-half inch asbestos packed iron stop cock on blow off pipe; one and one-quarter inch Pratt & Cady horizontal swing check valve for boiler feed.

All valves other than the five inch angle valves to be brass bodied; the water column to be attached to boiler, pipe at the bottom to be one and one-quarter inch, and at the top to be one inch, all bends must be made with Malleable iron or crosses to facilitate cleaning, and all spare openings fitted with square-headed brass plugs.

With each boiler there shall be furnished in addition to the foregoing, a full and complete set of fire and cleaning tools, including slicing bar, poker, coal and ash hose, steel scoop, and steam soot blower with proper attachments complete. Contractor will also furnish a complete set of grate and bearing bars, together with necessary buck staves and anchor rods, all in place.

All necessary tube and cleaning doors, wrought iron ladders and platform to be furnished and erected complete.

PRESSURE.—The boilers shall be designed to withstand a working pressure of one-hundred and twenty-five pounds per square inch. No cast iron will be allowed in the construction other than that for hand-hole plates, flanges or nozzles.

SPECIFICATIONS.—In making his proposition, the bidder will give full and complete specifications for the boilers he purposes to furnish.

GUARANTEES.—The contractor shall make the following guarantee for each boiler, which he is to furnish:

(a) That it will evaporate thirty-nine hundred pounds of water per hour from one hundred degrees into steam at seventy pounds pressure, or its equivalent.

(b) That in case of necessity it will evaporate fifty-eight hundred and fifty pounds of water under above temperature and pressure.

(c) That under guarantee (a) the steam will not contain more than ...... per cent entrainment.

(d) That when working at twenty-five per cent. over guarantee (a) the steam will not show over ...... per cent. entrainment.

(e) That when working under guarantee (a) it will show an evaporation from and at two hundred and twelve degrees of .... pounds of water per pound of coal. (The coal used to be what is shown as "Indiana block.")

(f) That when working at twenty-five per cent. over guarantee (a) it will show an evaporation from and at two hundred and twelve degrees of .... pounds of water per pound of coal. (Same coal as above specified.)

(g) The maximum allowable pressure which may be carried with safety.

(h) That the material, workmanship and finish shall be of the best throughout, and to replace and make good any parts which may prove defective, due to faulty workmanship or material, when operating under normal conditions, within one year from the final acceptance of the plant, without cost to the County.

BREECHING.—The contractor shall furnish and install complete, ready for service, smoke connections, all made of number twelve iron, the area to be sufficient to meet all requirements; the smoke connections to be provided with clean-out doors and hand dampers. The dampers will each be operated by suitable hand attachments easily accessible at front of boilers. Breeching to be securely fastened in position by wrought iron hangers, and to extend from the rear of the boilers to a common connection leading to the chimney located substantially as shown on the drawings. A design of breeching to be submitted by contractor before work is commenced, for the Commissioners' and Architect's approval. The whole to present a neat and finished appearance when completed.

DAMPER REGULATOR.—Contractor will furnish and install in a suitable location, one damper regulator of the "Spencer," "Clevauc" or other make equally good and acceptable to the Commissioners and Architect. Regulator to be complete with all attachments and properly connected to the damper to be provided and put in place in the main breeching leading to chimney, all to be complete and ready for operation.

ASLH PANS.—With each boiler there will be provided and put in place a wrought iron ash pan fitted with adjustable sliding damper so arranged as to permit of entirely closing off the opening to ash pit below, when burning natural gas. Pans to be paved with hard burned brick.

BOILER SETTING.—The contractor shall furnish all labor and material necessary for the proper enclosing of the three boilers, as specified, in one battery in the boiler room of the Power Station; boilers to be located substantially as shown on the drawings. All walls to rest on foundations specified under "Foundations" heading, said foundations to come within twelve inches of the finished boiler room floor, this specification covering all masonry work above such point. The entire inside walls directly exposed to heat to be lined with the best quality "Franklin Crown," or other fire brick equally good and acceptable to the Commissioners and Architect; all fire brick to be laid flat with headers every fourth course, the upper course of headers to be not lower than two courses below the lower row of tubes, all fire brick to be laid and rubbed into place with fire clay; all other than fire brick to be selected, first quality, hard burned brick; all outside walls, however, to be faced with first quality white enamel brick, same to be selected by the Commissioners and Architect. All outside walls to be protected by at least a two inch air space.

Contractor to set all ash and cleaning doors complete. All the work to be done in a workmanlike manner, and to present a neat and finished appearance when completed.

The contractor guarantees that all the material used will be the best of its respective kind that can be obtained.

PAINTING.—After the completion of his work, contractor will paint fronts, buck staves, cleaning doors, etc., with three coats of paint, thoroughly rubbing down the first two coats; the color to be such as may be selected by the Commissioners and Architect.

INSURANCE.—Each boiler will be inspected and insured by a responsible steam boiler insurance company, and a certificate of inspection and insurance policy for one year shall be furnished by the contractor; the amount of insurance to be placed on each boiler to be Five Hundred Dollars; policies to be fully paid by the contractor.

GAS PIPING AND BURNERS.—Contractor will run a six inch pipe from the curb line leaving same ready for gas company to make connections thereto. This pipe to run along the south side of the Power Station, entering the building at a point opposite the front of the battery of boilers. Where pipe runs in boiler room same will be laid in the trench elsewhere specified, connections to be made by double-sweep tees. to the six inch main at each boiler for pipes leading to each of the fire chambers. These pipes will be sufficient in number to supply the required amount of gas, each pipe being provided with the necessary number of gas burners for obtaining the results guaranteed by the contractor for each of his boilers. Either the "Whysall" or a cast iron burner as made by the Standard Oil Company, may be acceptable.

Contractor to make all connections, with the exception of that to the street main, complete and ready for operation, providing suitable valves on the main six inch intake pipe and in each of the several pipes leading underneath the boilers. All piping to be perfectly tight, the whole to be done in a workmanlike manner and to present a neat and finished appearance when completed.

All the piping to be so arranged and installed as to permit of the grate bars being readily placed in position or removed without necessitating the disturbance of any of the gas piping, and should occasion arise to permit of using the natural gas at the same time as when burning coal.

The contractor will also furnish and provide a moveable cast iron cover for the trench as above specified.

· ENGINES.

NUMBER AND TYPE.—The contractor shall furnish and erect complete in all details ready for making connection to piping and deliver in good working order, three simple, automatic cut-off, horizontal, high speed engines of the "McIntosh-Seymour," "New York Safety," (straight-line pattern), or other make equally good and acceptable to the Commissioners and Architect, suitable for direct connection to dynamos The engines to be of the latest and most approved pattern, extra heavy frames and fly-wheels and especially adapted for the service contemplated; all three of the engines to be right hand.

In case only one dynamo is coupled to an engine, then either a side crank or a center crank engine will be applicable, but should a three wire system of electrical distribution be adopted where it will be necessary to couple two units to the one engine, then only center crank engines will be considered. For center crank engines in any case, the crank shaft must be forged in one solid piece.

RATING.—Each engine at a speed of two hundred and fifty revolutions per minute shall develop from one hundred and thirty-five to one hundred and fifty indicated horse power, when cutting off at one-quarter stroke, with initial steam pressure of ninety pounds. In the selection of these

engines, the contractor must bear in mind that in addition
to the lighting, current will be supplied for electric elevator
service.

FITTINGS.—Each engine shall be provided with the following:

One throttle valve fitted to steam chest; one exhaust valve
(straightway); one automatic sight feed cylinder lubricator;
a full set of graduating oil cups; one set of wrenches, polished brass drip collectors and splashers; one full set of
brass hand-oilers, with tray.

All necessary pipes, valves and attachments permanently
connected with the engine for taking indicator cards from
both ends of the cylinder, connections to be fitted with a
three-way cock, pipe and fittings to be nickel plated.

A suitable reducing motion device shall also be furnished,
said device to be fitted with an arc and so arranged that it
can readily be attached to or detached from the engine frame.

A full set of automatic relief and drain valves, together
with all requisite drain pipes fitted to steam chest, ends of
cylinder and throttle valve; one iron sub-base of such design
as will mount the engine and dynamo.

GUARANTEES.—The contractor hereby guarantees,
together with what is required in the "General Conditions."

(1.) That engines shall operate noiselessly and without
vibration.

(2.) That all parts shall be carefully made to standard
guage; all moving parts carefully balanced and all valves
and packing free from leakage.

(3.) That any engine shall be capable of being operated
continuously for twenty-four hours under any load up to its
maximum rated capacity without undue heating of journals
or bearings.

(4.) That the material and workmanship shall be the
very best throughout.

(5.) To hold himself responsible for faulty workmanship
and material and to replace and make good any defects
which may develop within one year from the successful
starting of engines, without cost to the County.

GUARANTEED DATA TO BE GIVEN BY CON-
TRACTOR.--

(1.) Diameter of Cylinder, ...

(2.) Length of Stroke, ...

(3.) Diameter of Steam Pipe, ...

(4.) Diameter of Exhaust Pipe, ...

(5.) Diameter and Face of Fly-wheel, ...

(6.) Weight of each Fly-wheel, giving number of wheels, ...

(7.) Weight of Engine without wheels, or sub-base, ...
....

(8.) Weight of sub-base, ...

(9.) Engine center or side crank, ...

(10.) Speed (revolution per minute.) ....

(11.) Point of cut-off for obtaining maximum economy,
....

(12.) Indicated H. P. obtained when operating at above point with ninety pounds initial steam pressure at above rated speed, ...

(13.) Indicated H. P. when running at no load, ...

(14.) Maximum indicated horse power obtained at latest point of cut off at which governor will regulate within two per cent; initial steam pressure ninety pounds at speed given as above given, ...

(15.) The steam consumption per indicated horse power per hour when running under normal and regular conditions with initial steam pressure ninety pounds, at one-quarter cut-off; steam not to contain more than one per cent. moisture.................. ....

(16.) The maximum speed variation (in revolutions) be-- tween no load and full load, as well as the regulation through- out all changes of load, steam pressure varying from a minimum of eighty-five to a maximum of ninety-five pounds,
....

(17.) The contractor shall specify the combined efficiency of engine and dynamo which he guarantees at full load, and is requested to state what combined efficiency he will guar- antee at half load. In calculating the combined efficiency, the engine manufacturer shall assume that all bearings are a part of the engines and that the friction of all bearings is in- cluded in the engine friction, and that the commercial effi- ciency of the dynamos is 91 per cent. at full load and 88 per cent. at half load, (full load being taken as 800 amperes and 117 volts and half load being taken as 400 amperes and 115 volts. ....

IN GENERAL.—Immediately after the plant is in suc- cessful running order the contractor shall fill and thoroughly rub down the engine frames and then have same painted in a neat manner, and striped with gold leaf; the color to be determined by the Commissioners and Architect.

In case the sub-base of the engine selected be not provided with an oil groove around base, other suitable provision shall be made in the way of base trough, which shall be fitted around the bottom of sub-base to prevent any oil from reaching the floor, trough to be so arranged that it can be readily drained.

In making his proposition the bidder must understand that he is to provide the necessary length of shafting together with outboard, self-oiling bearings complete, for the armatures of dynamo machines, all as may be required.

DRAWINGS.—Complete outline drawings of engines and details of foundations must be furnished the Commissioners and Architect immediately after the selection of engines.

ERECTION AND STARTING.—Contractor shall furnish an expert to erect, adjust, start and run the engines for one week or until such time thereafter as may be necessary to put the engines in successful working condition. The expert to instruct the County's engineer as to the proper method of operating engines or any other apparatus, which he may furnish, under these specifications.

BOILER FEED PUMPS.

The contractor will furnish two boiler feed pumps, together with all necessary fittings, connections and appliances as hereinafter set forth. The apparatus to be delivered in such condition as to readily be placed in position and connected to piping

TYPE AND CAPACITY.—The contractor to furnish complete with all necessary fittings two boiler feed pumps each to be four and one-half by two and three-quarters by four inches, of the "Worthington," "Snow," or other manufacture equally good and acceptable to the Commissioners and Architect; pumps to be of the Duplex Plunger pattern suitable for pumping hot or cold water. With each of these pumps there will be furnished one cast iron table or stand having a drip pan in combination therewith, together with all necessary bolts and nuts for properly fitting up.

FITTINGS.—Each pump will be provided with the following: Air and drain cocks, one half pint Detroit lubricator, or one equally good; complete set of oil cups; cylinder blow globe valves permanently attached to steam cylinders.

GUARANTEE.—Contractor guarantees the pumps which he furnishes to be of the best workmanship and material throughout, and that the same will have a capacity of pumping at least twenty gallons of water per minute without excessive piston or overtaxing its capacity in any way.

Contractor further guarantees to replace and make good any parts which develop defects due to faulty workmanship or material within one year from date of final acceptance of his apparatus, without cost to the County.

INJECTORS.

The contractor will furnish complete, ready for connections, two "Sellers" re-starting injectors each capable of delivering at least one thousand gallons of water per hour with a minimum steam pressure of eighty pounds.

Contractor guarantees to replace and make good any parts which may develop defects within one year from date of final acceptance of his apparatus, without cost to the County.

SEPARATORS.

The contractor will furnish complete with all attachments, ready for making connections, three verticle six inch steam separators suitable for attaching direct to throttle valve of engines, and three horizontal seven inch oil separators for exhaust piping; separators to be the "Zig-Zag," or other make equally good and acceptable to the Commissioners and Architect. Each separator will be provided with automatic, self-closing water gauge cocks.

Contractor guarantees the separators as furnished by him to be of the best material and workmanship throughout, and to hold himself responsible for faulty construction and to replace and make good any part which may develop defects due to faulty workmanship or material within one year from the date of final acceptance of his apparatus, without cost to the County.

STEAM TRAP.

The contractor will furnish complete and ready for making connections one number four "Acme" steam trap, or other make equally good and acceptable to the Commissioners and Architect; trap to be provided with automatic, self-closing water gauge cocks.

EXHAUST HEAD.

The contractor will furnish complete, ready for attaching to exhaust pipe, one ten inch "Lyman" exhaust head, or other make equally good and acceptable to the Commissioners and Architect.

HEATER.

The contractor will furnish complete in all details one exhaust steam open feed water heater of the "Webster," "Cochrane," or other make equally good and acceptable to the Commissioners and Architect; heater to have a capacity of five hundred horse-power.

Heater will be provided with an automatic intake valve and be furnished with suitable water guage.

The heater as furnished is to be used as a receiving tank in connection with the steam heating system.

The contractor guarantees the heater which he furnishes to be free from all mechanical defects, and that the best workmanship and material has been used in its manufacture; he further guarantees to replace and make good any defects which may develop due to faulty construction or material within one year from the date of the final acceptance of his apparatus, without cost to the County.

GUARANTEES.

All guarantees made to be subject to and in connection with the "General Requirements."

PIPING OF POWER PLANT.

GENERAL.—The contractor is to furnish all piping, valves, fittings and accessories necessarily involved. and to install complete in all details the entire system of Steam and Exhaust piping, together with water connections from both City pressure mains and well in jail yard to boilers. pumps and heater. It is intended that the following shall cover completely and entirely the whole work involved in a plant such as is contemplated, and in case any part of these specifications should not be perfectly clear to the contractor, he is requested to immediately communicate with the Architect for such explanation as may be required, as that, after the execution of the contract, a complete and perfect system will be expected and demanded of the contractor.

In his proposal, the bidder is requested to specify the make of all fittings and accessories he proposes to install, excepting where special types are already called for in these specifications.

PIPING.—All pipe used to be standard, full weight, wrought iron, lap welded pipe throughout, with flanged fittings for sizes above three inches. Pipe guaranteed to withstand a working pressure of one hundred and fifty pounds per square inch.

"Rainbow" or "McKim" gaskets of first quality to be used throughout.

Contractor shall connect each of the main five inch angle valves, as elsewhere specified, to the boiler nozzles, and connect same by five inch pipe to an eight inch main steam header, pipe to rise from the boilers, run over and into the top of header which will be located above and toward the front of the battery of boilers, substantially as shown in the drawing. Header to be eight inches diameter, to be fitted with three, eight inch by eight inch by five inch double sweep tees. One end of header to be closed by blind flanges. The five inch pipes from boilers to pitch toward the header. The steam pipes for each of the Engines to be six inch pipe led out of the bottom of the main steam pipe by long bend fittings and connected to a main six inch "Chapman" valve, and then to special six inch copper bends of thirty inch radius, connecting to the vertical separators as elsewhere specified. Separators to be fastened directly to throttle valves of engines.

The contractor will furnish one four inch "Clevauc" pressure reducing valve, capable of delivering any pressure between ten pounds and a vacum equivalent to two pounds.

A six inch pipe will connect the steam header to the steam heating system, and in the pipe there will be placed a six inch valve which will be by-passed by a four inch pipe connecting with the pressure reducing valve, on either side of which will be placed a four inch straightway valve.

The one and one-fourth inch steam pipe for pumps to be taken directly from the main steam header by tapping the top of the pipe at a point most convenient to the pumps, which will be located substantially as shown in drawing; pipe to be carried to the pumps and connected by a double elbow to each pump. The main exhaust piping from each engine to be seven inches in diameter, located in position substantially as shown on the drawing; this exhaust piping to be "light" wrought iron, lap welded pipe with flanged fittings. Piping to run from each engine to a twelve inch exhaust header running to the feed water heater. The contractor to place an oil separator, as indicated, in each exhaust pipe; separators to be provided as elsewhere specified.

All drips from engines, exhaust header and oil separators to be connected by a one and one-half inch pipe to be carried to the catch basin located in the boiler room substantially as shown. All this piping to be suitably valved, using straightway valves.

The contractor to set up and properly connect the steam trap provided, as elsewhere specified; trap to be connected to each of the steam separators, with a check valve placed in each pipe. All connections to trap to be suitably valved and trap to be by-passed to catch basin. Trap to be arranged to discharge directly into feeder water heater, with by-pass connections to catch basins.

The contractor will furnish and make all necessary connections thereto, one automatic drainer or syphon as manufactured by Philip Braender, New York, or other make equally good and acceptable to the Commissioners and Architect. The drainer to be placed in the sump located at or near the entrance to the tunnel and all connection made to piping.

The heater, as elsewhere specified, is to be installed and all connections made to same as hereinafter set forth. From the heater a ten inch "light" wrought iron lap welded pipe is to be run up to and above the highest point of the roof of the building and there to be fitted with an exhaust head as elsewhere specified; the drip from exhaust head to be carried down to the catch basin. The contractor to make connections from exhaust piping direct to heater and by-pass heater to atmosphere, providing suitable valves as required, connections to be made from exhaust header to the fourteen inch heating main leading to the Court House, and to the four inch main leading to the Jail building, the contractor providing a straightway valve for the Court House main. The contractor to furnish and connect to the exhaust pipe leading to atmosphere one "Davis" noiseless back pressure valve, or other make equally good and acceptable to the Commissioners and Architect.

The two boiler feed pumps, as specified elsewhere are to be installed upon foundations as specified, and all connections to be made as called for herein.

The one inch exhaust from each of the boiler feed pumps to connect by a double elbow to a one and one-half inch pipe to be connected directly with the heater exhaust intake; all bends being made with water ells.

The contractor to attach the two inch pressure feed to City main, running same to interior of boiler room, as shown in the drawing, and connect same to heater intake valve, to injectors, to boiler feed pipes, and to main two inch pump suction. Piping to be suitably valved and arranged so that any connection may be used independently of the others or in conjunction therewith.

The suction of pumps to be taken directly from the heater by a three inch pipe; this pipe to be connected to the two inch pressure main as described, and to each two inch pump suction each fitted with straightway valve. The one and one-half inch discharge from each pump to be connected directly to each of the three boilers, a horizontal swing check valve to be placed in each of these pipes and one in each separate feed pipe to boilers, besides one back of each injector connection as shown in the drawing, all to be fitted with suitable straightway valves. The valves for each of the separate boiler feed pipes to be furnished with extension stems so that valves may be operated conveniently from the floor.

The two and one-half inch boiler blow-off and the one and one-half inch heater drip and overflow pipe to be carried to the catch basin direct. The stop cocks for the boiler blow-offs to be furnished as specified elsewhere.

The three and one-half inch safety valves, as elsewhere specified, will be attached to their respective boilers, and the four inch side openings from each will be connected to a six inch header at rear of battery and this pipe carried through the rear wall to outside of building.

The contractor to connect up complete in all details two injectors as provided. Steam connections to be taken directly from main steam header; water from pressure feed; the discharge being led and connected to each of the one and one-fourth inch boiler feed pipes, all to be suitably valved throughout; check valves placed as shown on drawings. The overflow from injectors to discharge into a copper funnel with a drip pipe carried back to catch basin. Injectors each to be placed at a convenient height for operating, and to be located substantially as shown on the drawings.

VALVES AND FITTINGS.—All valves used in the main high pressure steam piping to be "extra heavy" gate valves; all valves used in exhaust pipes, or in heating mains to be standard gate valves; all water valves to be "straight-way;" all valves as used throughout to be the best of their respective kind; steam and water valves up to and including three inch valves to be brass-bodied. Wherever globe valves are used, the same will be brass-bodied renewable seat valves, either the "Crane" or "Jenkins" seat. Each of the valves placed in the steam pipes leading to the separate engines to be by-passed.

All check valves to be brass-bodied, "Pratt & Cady", horizontal swing checks.

The contractor to connect to this system of piping the throttle valves, exhaust valves, etc., and provide all bolts, nuts and gaskets.

All flanges to be standard size and to be fitted with full sized "Rainbow" or McKim gaskets. All piping under three inch will be connected up with a suitable number of unions to permit of its being readily taken down for repairs or renewal.

HANGERS AND BRACKETS.—The main steam header shall be supported by substantial cast iron brackets furnished and securely fastened in position, brackets to be so placed as to firmly hold the header in position and at the same time permit of sufficient freedom for expansion and contraction. All other overhead steam or water piping, of whatever description, to be supported in position by suitable expansion pipe hangers spaced not more than ten feet centers. Where pipe is not supported by hangers same to be held in position by suitable hooks. All supports of whatever kind to present a neat and finished appearance when in position.

COVERING.—All steam pipes, valves and fittings, whether for live or exhaust steam throughout this plant, including exhaust pipe to the roof, shall be thoroughly covered with a one-inch "Magnesia," "Asbestos Sponge" or

"Nonpareil" cork, sectional covering. All covering to be carefully put in place, using brass bands; all seams to be perfectly tight; the whole to be painted with at least two coats of an air tight paint of such color as the Commissioners and Architect may select, and to present a neat and finished appearance when completed.

In addition to the above, the contractor will cover the feed water heater with a plastic, non-combustible, non-conducting covering at least two inches in thickness.

GAUGE BOARD AND GAUGES.— There will be furnished and installed a finished black enameled slate Gauge Board, to be not less than one and one-fourth inches in thickness, and to have beveled edges. Board to be of such size as not to crowd the instruments which will be arranged so as to permit of at least a six inch border all around. The board to be placed in the engine room at a convenient point substantially as shown on the drawings. The contractor will furnish and attach to this board complete, with all connections, drips, valves, etc., ready for service, the following appliances:

One eight-day clock; one high pressure "Crosby" guage for steam header, to register from one to one hundred and fifty pounds with proper sub-divisions. One "Crosby" compound pressure and vacuum gauge, to register from o to twenty pounds pressure and from o to ten pounds vacuum. One "Crosby" pressure-recording guage to register to one hundred and fifty pounds.

All the above to have eight inch dials with black faces and white figures; cases to be iron with nickel-plated rings. There will be engraved on each dial the following: "Power Plant Allen County Court House."

The steam and recording gauges to be connected to the main steam header by a three-eighth inch pipe in each case. Pipes to be suitably valved and to be so connected that the gauges may indicate the true pressure unaffected by any water column. The vacuum gauge to be connected to the steam heating system by a three-eighth inch pipe.

The contractor must guarantee in connection with the "General Requirements," all workmanship and material furnished under his contract to be the best throughout, and to replace and make good any defects which may develop in any part of the entire system which he installs within one year of date of final acceptance of his work, without cost to the County.

DYNAMOS.

NUMBER AND TYPE.—The contractor is to furnish and install complete in all details ready for service, three slow speed, mulipolar, compound wound dynamos, direct coupled to three engines, described in Engine specifications.

Each dynamo to have a capacity of eight hundred amperes at an E. M. F. of one hundred and thirty volts, and at a speed of two hundred and fifty revolutions per minute. Dynamos to be overcompounded five volts. It is the intention that each dynamo shall run two hundred and fifty revolutions per minute and that at this speed it shall have an E. M. F. of from one hundred and twelve to one hundred and seventeen volts; the regulation being accomplished by the compound field winding; but when two or three of the dynamos are running in multiple-arc, additional voltage will be required to overcome line loss and this is to be obtained by cutting resistance out of the shunt fields. The dynamos must, therefore, be capable of giving at least one hundred and thirty volts at normal speed when required. The exact ampere capacity above specified is not arbitrary, but any bidder offering dynamos which have an ampere capacity any less than above specified shall explicitly state the ampere capacity of the dynamos offered in his proposition. The dynamos must give a voltage equal to that specified at normal speed and at any temperature which they may reach during continuous operation at full specified load.

REGULATION.—Each dynamo to be automatic in its regulation from one hundred and twelve to one hundred and seventeen volts, so that any changes in load up to the specified capacity of the dynamo will not demand any special attention from the operator. The compound winding to be so proportioned that if the resistance of the shunt field is so adjusted as to give one hundred and twelve volts at no load, the voltage will be one hundred and seventeen volts at full load and not over one hundred and fifteen volts at half load, without any change of resistance in the circuit of the shunt field, provided the speed remains constant at two hundred and fifty revolutions per minute.

ARMATURES.—The dynamos to be of a standard make

and of the highest grade of design, workmanship and finish. The armatures to be of the type known as iron-clad. The conductors to have a layer of mica of substantial thickness, insulating them from the core, and additional insulation if necessary to fix them rigidly in place and to form a good mechanical separation from the core. The armatures to be perfectly balanced both electrically and mechanically.

COMMUTATORS.—All commutators to be preferably of hard drawn copper, and to be of such size as to fill the requirements of these specifications regarding heating and sparking.

BRUSHES.—All dynamos to be equipped with carbon brushes of such size as to provide a circuit for the current, having a cross section of not less than one square inch for thirty amperes of output of the machines up to the specified ampere capacity of the machine.

RHEOSTATS.—There shall be provided with each dynamo a Field Rheostat of approved design and suitable for controlling the dynamo in the manner indicated in this specification. Rheostat to be of suitable design to be placed below horizontal marble switch board and to be operated by polished copper wheel above the board; wheel to be equipped with a pointer to indicate position of rheostat switch.

BEARINGS.—Each dynamo to be provided with approved self oiling bearings and set up and connected to the engine complete and ready for service.

TRIMMINGS.—There shall be provided as a fixture to each dynamo a polished copper four light cluster fixture stand, complete with sixteen candle power, one hundred and fifteen volt lamps, keyless sockets, canopy, shade holders and fluted opal shades. The height over all not to exceed ten inches.

PAINTING.—Immediately after the plant is in successful running order the contractor shall fill and thoroughly rub down the dynamo castings and then have same painted in a neat manner, and striped with gold leaf; the color to be determined by the Architect.

EXPERT SERVICES.—The contractor shall furnish an expert to adjust, start and run the dynamos for one week, or until such time thereafter as may be necessary to put the machine in successful working condition. The expert to instruct the County's operator as to the proper method of operating the dynamos or any other apparatus or appliances which may be furnished under his contract.

GUARANTEES.—The contractor hereby guarantees in connection with the requirements of the "General Conditions:"

1. That each dynamo will carry its rated load in amperes at an E. M. F. of one hundred and thirty volts continuously for a period of ten hours and that when so operated the temperature of no part of the armature or fields shall rise to more than seventy-five degrees above the temperature of the surrounding air and the temperature of the commutator shall not rise to more than one hundred degrees Fahrenheit above that of the surrounding air.

2. That each dynamo will automatically regulate its voltage as specified in above article entitled "Regulation."

3. That all armatures shall be perfectly balanced and smooth running.

4. That each dynamo will run continuously with entire freedom from sparking at brushes under any condition or change of load from no load to full load without shifting of brushes, and that each machine can, if it is desired, be operated at twenty per cent. overload continuously without any sparking at the brushes (with shifting of brushes if this be necessary).

5. That the insulation resistance between any conductor upon either field or armature and any part of the iron work of any of the dynamos shall not be less than two meghoms.

6. That each dynamo shall have a commercial efficiency of not less than ninety-one per cent. at full load in amperes and 117 volts and not less than eighty-eight per cent. efficiency at half rated load in amperes and 115 volts.

7. That each dynamo will, if required, carry a load of twenty per cent. in excess of the specified ampere capacity for at least four hours without any injurious overheating.

8. That the dynamos will operate and regulate as successfully when two or all of them are connected in multiple-arc as when one is being operated alone.

That each dynamo shall be mechanically and electrically perfect in workmanship and of the best material throughout, and to replace and make good, without expense to the County, any defects due to imperfect workmanship or material that may develop within one year from the date that the dynamos are first put in successful operation, provided the dynamos shall be operated with proper and ordinary care under the conditions of operation indicated in these specifications.

ELECTRICAL SYSTEM.—All' electrical specifications are drawn up on the basis of installing machinery and circuits to operate all lights and motors by DIRECT current and at a pressure of one hundred and ten volts. In drawing up the specifications in this manner it is not the intention to prevent the use of any other system than the one specified, but it is intended that no other system shall be substituted for the one specified unless it shall, in the opinion of the Commissioners and Architect, be as reliable, as economical to operate and in every way as desirable as the system specified. If any other system is furnished, specifications for dynamos, switch board, wiring in Tunnel and wiring of Court House will be revised, but the present specifications shall be followed as far as they apply to the system adopted, and any changes to be made in the specifications shall be submitted by the contractor and shall be approved in writing by the Commissioners and Architect before construction commences.

A three wire system will be accepted, provided the saving in cost is, in the opinion of the Commissioners and Architect, sufficient to justify the use of additional machinery and devices.

If a three wire system is used then all conductors shall be figured for substantially the same per cent. loss as in the present specifications, (except as otherwise specified in specifications for "wiring in Tunnel,") all circuits except the main feeder from Power Station to Court House shall have positive, negative and neutral conductors of the same size; all motors will be run on outside wires at two hundred and twenty volts; each engine may be provided with a pair of dynamos and the switch board will be equipped with the necessary instruments and devices, so that the plant can in every way be operated the same as if it were erected according to the present specifications.

The contractor may (and it is desired that he will) submit a proposition upon a one hundred and ten volt system as specified, whether he bids upon some other system or not, but he must submit at least one bid upon a direct current two wire or three wire system, using one hundred and ten volt lamps. In case any bidder wishes to submit a prop-

osition upon any system, not above mentioned, he may
also submit a proposition for such a system, but the burden
of proof shall be on him to demonstrate that such a system
is as good as the one specified, and unless he shall submit
such a demonstration, along with his proposition, the proposition will not be considered.

SWITCH BOARD FOR POWER STATION.

The Switch Board shall be located substantially as shown
on plan of station, and shall be of the form shown on drawing, which is hereby made a part of this specification.

The face of the board will consist of two parts; a slab or
table slightly inclined from the horizontal, which will carry
all switches and rheostat wheels and a verticle slab, which
will carry all instruments, also brackets for lamps to illu-
minate the board.

All conductors to be run beneath the floor and to be run
up to and down from the switchboard in the space below the
horizontal part of the board.

IN GENERAL.—This specification includes the switchboard and all necessary switches, instruments, conductors
and controlling and regulating devices mounted upon the
same ready for use, excepting rheostats, which are in the
specifications for dynamos, but this specification includes
the mounting of rheostats on switchboard, and the furnishing and installing of all necessary conductors and connections to the same.

This specification is intended to include all labor and material that is a necessary part of or necessary to a complete
switch board, completely equipped and all connections between dynamos and switchboard and between circuits and
switchboards, (except the rheostats above mentioned.)
There is to be no metal in sight upon the face of the horizontal portion of the switchboard except polished copper.

CONSTRUCTION OF BOARD.—The Switch Board
will be constructed of black enameled slate of high insulating qualties, the horizontal portion of the board to be of
one slab of not less than one and one-quarter inches in
thickness, each of these slabs to be three feet wide and as
long as is necessary to accommodate the instruments, preferably not longer than six feet.

The board to be surrounded on the front and sides with

a base board of the same material as the board, eight inches wide and one inch thick. The verticle portion of the board to have a top and sides of the same material as the board, six inches wide and one inch thick, the side pieces extending down to the base board. The open space thus left on the front and sides of the board to be filled with grating or open work metal of design and finish acceptable to the Commissioners and Architect. The back of the board to consist of two iron doors hung from hinges at the ends of the board. The entire board to be rigidly and substantially supported by angle irons concealed within the board, and wherever the angle irons are fastened to the slate, they are to be held in place by bolts or studs, having polished copper heads.

INSTRUMENTS.—There will be furnished and erected in place:

One Weston Illuminated Dial Ammeter, with scale readone hundred and forty volts.

Three Weston Illuminated Dial Ammeters, with scales reading up to one thousand amperes.

One Weston Illuminated Dial Ammeter, with scale reading up to three thousand amperes.

One Multiple Arcing Galvanometer, of approved type, for indicating when the E. M. F. of any dynamo is the same as the E. M. F. of the Bus Bars.

One Ground Detector, of approved type.

SWITCHES.--The following switches to be furnished and connected in place upon the horizontal portion of the board:

Three eight hundred ampere three pole switches, (one for each dynamo.)

Two two thousand four hundred ampere single pole knife switches, one for each pole of main feeder.

One one hundred ampere double pole knife switch (for circuit to Sheriff's House and Jail.)

One three circuit volt meter switch (for connecting the volt meter to either bus, pressure wires, or sub feeder.)

One Multiple point switch (for shifting multiple-arcing Galvanometer from one set of dynamo leads to another.)

All ampere meters to be of shunt type, provided with connections of such lengths that the shunts can be placed below the horizontal portion of the board, and so located as to introduce no resistance into the circuit except the resistance of the shunts themselves.

If the ground detector consists of the ordinary form of incandescent lamps and push button, then the lamps shall be small ten c. p. lamps, the push button shall be of flush type and placed on the horizontal portion of the board.

All knife switches to be of the make known as the Chicago Edison Company's switch, constructed of pure hard drawn copper, and having surface contacts of not less than one square inch for every fifty amperes of current, and a sectional area in blades and clips of not less than one square inch for every one thousand amperes of current.

All knife switches to be equipped with safety strip holders, and to be so placed that in opening the switch, the motion will be away from the front of the board.

BUS BARS.—There will be three bus bars, one positive, one negative and one equalizing bus; each to have a cross section of not less than two square inches, and to be so proportioned that the current density will not be greater than eight hundred amperes per square inch in any part of any bar.

WIRES AND CABLES.—The following wires and cables to be furnished, connected up in place with all necessary terminals and connectors ready for operation:

One pair of main dynamo leads from each dynamo, each lead to have a sectional area of not less than one million circular mils.

One equalizing lead from each dynamo, each lead having a sectional area of not less than one million five hundred thousand circular mils.

All dynamo leads to be of stranded copper wire of not less than ninety-eight per cent. conductivity, to be rubber covered and leaded, and to have a covering of compounded jute outside the lead for mechanical protection.

Cables to be furnished in place and connected up to their respective dynamos and switch boards ready for use.

The positive and negative dynamo leads to be so proportioned and adjusted that the resistance of each pair from the dynamo to the bus bar shall be the same within five per cent.

The equalizing leads to be so proportioned and adjusted

that each lead shall have the same resistance from dynamo to bus bar within five per cent.

There shall also be furnished leads from the dynamos to their field rheostats, of such sizes and so connected as may be required by the manufacturer of the dynamos.

The pressure lines are included in specifications for wiring in tunnel, but the connecting of pressure line and all wiring and connections to volt meter switch and galvanometer switch are included in this specification.

The furnishing of main feeder is included in the specifications for wiring in tunnel, and all wiring except the wiring on switch board and between the dynamos and switch board is included in other specifications, but this specification includes the connection of the main feeder and all other lighting circuits to the switches above mentioned.

All wires and cables from dynamos to switchboard to be run either in vitrified tile duct or upon non-combustible insulators, if accessible.

CONNECTIONS.—There shall be furnished all necessary connections, connectors, cable terminals, etc., for connecting all wires and cables on switchboard and for making all connections to dynamos. All connectors, terminals, and in general, all conductors of large currents of which the size has not been specified above, to be designed upon the basis of a current density of not more than one thousand amperes per square inch of pure copper, and when any metal of less conductivity is used, the cross section shall be correspondingly increased; their design shall also provide a surface contact of not less than one square inch for every one hundred amperes where bolted or clamped connections are made.

BRACKETS.—There will be furnished two two-light, polished copper brackets (complete with sockets and shades of the same finish, and four sixteen c. p. lamps) for illuminating the horizontal portion of the board.

RHEOSTATS.—The rheostats are included in dynamo specifications, but their mounting and connecting is included in this specification; they must be so connected that a right hand rotation of the wheel will cut resistance out of the field circuit.

PLANS.—Before starting in work upon the switchboard, the contractor shall submit to the Commissioners and the Architect for their approval, complete detail plans, together with description of everything which he proposes to furnish under this specification; and work shall not be taken up until such plans and descriptions have been approved in writing by the Commissioners and Architect.

The Commssioners and Architect shall have the right to change arrangement and dimensions of any part of the switchboard at any time before plans are approved, and such change shall not be made the basis of any extra charge on the part of the contractor.

Should the electrical system adopted be other than a two wire system, using one hundred and ten volt lamps and motors, then the switch board will be changed to correspond to the system adopted, and in such event, the contractor shall, before starting in construction, submit for approval of the Commissioners and Architect, complete detailed plans and specifications of what he proposes to furnish. The above specifications shall be made the basis of his specifications, and the design and material shall be equal in every way to that herein specified. Written approval of such plans and specifications shall be received from the Commissioners and Architect before construction is commenced.

Whatever system is adopted, the losses of energy in dynamo leads and upon switch board shall not be in excess of what they would be if the plant were installed as called

WIRING IN TUNNEL.

The wiring in the tunnel will comprise all the circuits between the switchboard in the Power Station and the center of distribution in the Court House, and this specification covers the furnishing of all such circuits complete, including wires, cables, cable heads, insulators and supports, except the pipe hooks mentioned in this specification and shown on plan of tunnel, which are included in the specifications for the tunnel.

There will be three circuits in all:

1st, a main feeder for carrying all the current generated in the station to the center of distribution in the Court House.

2nd, a sub-feeder for carrying current, for operating lights in Power Station, Jail and Sheriff's House, from the center of distribution in the Court House, back to the switchboard in the Power Station.

3rd, a pressure line from the center of distribution in the Court House to the volt meter located on the switchboard in the Power Station.

MAIN FEEDER.—The main feeder shall run from the switchboard in the Power Station to the cut-out board in the basement of the Court House. Each pole of the feeder shall have a sectional area of not less than five million circular mils, or to a total of ten million circular mils in both conductors. The feeder may be made up of two cables or a number of cables connected up so as to be equivalent to five million circular mil cables, but no cable to be used of a size less than one million circular mils. The cables to be of soft drawn copper wire of not less than ninety-eight per cent conductivity, and to be insulated with a solid paper insulation, the paper being thoroughly impregnated with some moisture proof compound. The paper insulaton to be surrounded with a continuous covering of lead, and the lead to be covered with an external coating of four-ply jute, thoroughly saturated with a waterproof compound to prevent mechanical or chemical injury to the lead.

SUB-FEEDER.—There shall also be a sub-feeder run from the cut-out board in the Court House to the switchboard in the Power Station. Each pole of the sub-feeder to consist of a cable of five hundred thousand circular mils; these cables to have the same kind of insulation as the main feeder, and the insulation to be protected by a lead covering, and the lead to be protected by a compound jute covering in the same manner as the cables of the main feeder.

INSULATION OF CABLES.—The paper insulation of all of the above cables to be three-sixteenths of an inch in thickness, and the lead to be three thirty-seconds of an inch in thickness.

The contractor to specifically state what insulation resistance he guarantees for each cable when the cable is in place and connected up to terminal heads.

PRESSURE WIRES.—There will be one pair of pressure wires run from the bus on the cut-out board in the Court House to the volt meter switch on switchboard in the Power Station. This circuit to be of number fourteen B. & S. guage wire, having a high grade of rubber insulation and protected with an external covering of lead, or a duplex wire with one covering of lead may be used, preferably the latter.

SUPPORTING OF WIRES AND CABLES.—All cables and wires to be supported upon the pipe hangers above mentioned, in the tunnel, unless other supports are needed, in which case, such other supports shall be supplied and erected by the contractor. Within the Power Station and the Court House, the cables to be run upon incombustible insulation of approved design, and the pressure wires to be run in an iron armored conduit.

TERMINALS.—There shall be provided for each end of each cable a suitable hard rubber terminal, which shall be placed upon the cables and compounded so as to absolutely exclude all moisture from the insulation of the cable. Care must be taken to see that these terminals are perfect after the copper terminals are connected to the cables.

JOINTS.—All wires and cables to be in as long lengths as can be secured from any responsible manufacturer, preferably to be in one piece. If there are any joints, they must be so made that the joints in both insulation lead and external covering shall as nearly as possible be equal to any other portion of the cable.

IN GENERAL.—The cables must be of ample length to form convenient connections at both ends, and they shall be made up of a sufficient number of strands, so that when installed they will present a neat mechanical appearance. A sample of each size of cable to be submitted to Commissioners and Architect for approval before construction commences.

Everything specified to be furnished complete and in place ready for use.

NOTE.—If any other system except a two wire system with one hundred and ten volt lamps and motors be used, then the size and number of conductors will be correspondingly changed. If a three wire system is used, then the main feeder shall consist of a positive and negative, each of one million five hundred thousand circular mils, and a neutral of five hundred thousand circular mils; and the sub-feeder shall consist of positive, negative and neutral, each of 4-0 B. & S. guage, and the pressure current will consist of three No. 14 B. & S. guage wires.

BRICK STACK.

Excavate to a point three feet below the present level of the ground.

Borings will be made to ascertain the character of the material underneath the point where stack will be located, and if in the judgment of the Architect, it is necessary to use piling, the contractor shall supply the same in place, and shall be paid extra for that, an amount to be agreed upon at the time.

All cement and concrete used in this work shall be as specified for the foundations for buildings.

Bricks for the outside surface must be culled and used so that the exterior of the stack will present a uniform color. Brick work to be as specified for the foundations for buildings.

All fire brick used in the connection from boiler house to stack flue, and extending up the latter to forty foot level, shall be those known as the "Franklin Crown" or other equally good and acceptable to the Commissioners and Architect. From the forty foot level up to the eighty foot level, a No. 2 quality may be used. All fire brick to be laid in fire clay with rubbed joints; no more clay being used than just sufficient to fill the unevenness of the bricks. All fire brick used in the core of the stack shall be curved to the proper radius.

Stone work to be as shown on details, all of Blue Bedford lime stone, or other stone acceptable to the Commissioners and Architect.

The connection between the boiler house and the stack flue will be of brick and iron as shown. It will be supported from two six inch wrought iron I beams, one end of each resting on the outside of the stack base. These I beams shall be five feet apart, center to center, and across them resting four inches on the boiler house wall; the other shall be placed on cast iron plates one inch thick and heavily ribbed. Upon these there shall be built two nine inch brick walls, the outside course of which shall be of red brick; the inside course of fire brick. These walls will extend up to a heighth of five feet eight inches, at which point cast iron cover plates shall be laid across to form a roof. These cover plates may be about twelve inches in width, and each shall have a rib running its whole length, and a lip covering the joint between it and the plate adjoining. Upon these cover plates there shall be laid one course of red brick, laid flat in mortar, and upon these again, one course of paving brick on edge raised at the center, so as to pitch both ways. The inside of the connection shall be paved with one course of red brick laid flat, and one course of fire brick laid edgewise.

The top of the brick cap of the main shaft of the stack will be covered with a cast iron shell, as shown on detail drawing. This shell shall be made in eight pieces, which shall be bolted together at the flanges, and the joints filled with some acceptable compound, so as to exclude moisture.

The top of the core cap will also be covered with a cast iron shell, the top of which shall be surmounted by a copper ring, all as shown on detail drawing. The cast iron shell is to be made in eight pieces, bolted together at flanges, and the copper ring shall be made with as few joints as possible, and shall be set so that its joints do not come over the joints of the iron shell below. The copper ring will be held down to the brick work and to the cast iron shell beneath by copper bolts as shown.

On top of the stone base course the brick work will commence. The flue in the center of the base will be circular, five feet nine inches in diameter. It shall be of red brick throughout from the top of the concrete foundation up to about one foot below where the connection to the boiler house enters the flue. Above this point it shall be lined with four inches of fire brick. Between the wall forming the flue and the main body of the brick work of the base, there shall be an air space four inches wide. This air space shall entirely surround the core, excepting at the point where the connection to the boiler room enters and at point where cleaning door is placed. The connection to the boiler room will be an arched opening of size and shape shown. The opening will be supported by three courses of brick placed edgewise, the inside course being of fire brick as above specified, and being joined into the brick work of flue connection already described, and also into the fire brick lining of the core. Where the flue connection enters the core at the top, the arch shall be slanted up, so as to give free egress to the hot gases from the connection into the core. The cleaning door will be placed on that side of the base opposite connection to boiler house. It will be a round top cast iron door and frame, two feet wide in the clear and about three feet high, and the opening for it shall be carried by three courses of red brick laid edgewise, the arch running through, into and connecting with the core. On the inside this opening is to be plugged with a dead wall one brick thick.

The space which is to be left in the center of the concrete foundation as shown on drawings, shall be filled with earth or cinders tightly rammed, and shall be paved with one course of paving brick on edge at the level of cleaning door. Air holes one course high and one and one-half inches wide shall be left in the lower part of the base about one foot from the ground level, extending into the air space surrounding the core. Additional air holes shall also be made through the outside core at sixty foot level.

Great care must be taken by the contractor throughout the whole construction of this stack that mortar or portions of brick are not allowed to drop into the air spaces and obstruct the same.

Directly under the stone cap, the brick work will be racked out for six courses, until it is fifteen feet square, at which dimension it will extend up plain one foot in height, forming a cornice as shown.

The main shaft of the stack will be circular at all points in cross section, and of the diameters figured on drawing.

The core of the shaft will commence at the top of the stone cap and will be of the diameters figured on drawing.

The inside course of the core will be fire brick of quality already described, up to the eighty foot level.

Both the external and internal diameters referred to above and figured on the drawings must be rigidly followed by the contractor.

Special means must be provided by the contractor whereby the work may be laid with the greatest accuracy. Great care must be taken at all times as already indicated to keep the air spaces clear so that no obstruction may be presented at any point to the free movement of the core due to hot gasses passing therein.

At points twenty-five feet, fifty feet, and one hundred feet above the cap stone, headers will be laid in the outside shell, projecting into the air spaces to within one-quarter inch of the outside face of the core at this point. These headers may be laid in a ring surrounding the core; the one-quarter space mentioned must be filled up with soft wood.

As the drawings show, the core will extend eighteen inches above the top of the shell, at which point it will be thickened up to one brick thick, and covered and protected by a shell of cast iron and a copper ring at the top. Great care must also be observed at this point to preserve the brick work already laid, and to see that everything is solid and substantial.

The copper ring shall be put on last and shall be laid on a thick bed of mortar, upon which it shall be screwed down tight by the bolts already described.

After completion, both the cast iron caps shall be painted with two or more coats of some weatherproof paint, of quality and color satisfactory to Architect.

A wrought iron ladder, built as per drawings, shall be furnished and put up on the side of the stack away from the boiler house, supported from the stack itself. The detail drawings show the shape and sizes of the supports and the method of attaching the ladder proper to them.

For protection from lightning the shaft cap will be surrounded by a copper ring, supported by eye bolts screwed into the cast iron cap. The ring must be free to move in the eye bolts, and from it shall extend upward; the points, in number and location as shown. From the ring, the main rod, in the form of a copper strip one-eighth inch by two inches, will extend down on the north side of the stack away from the boiler house, supported from it in the manner shewn by the drawings. The points shall extend up above the core caps through sleeves, so that the connection shall be made flexible, and in such a way that any movement of the core shall not destroy, or in any way, effect the rod itself. At the ground end, the rod shall be run in a trench to a point to be selected, about fifty feet away from the building, at which point it shall terminate in a copper or iron plate approximately two feet square to form a ground connection. The grounding plate to be well imbedded in broken coke and buried in moist earth below the level of the river.

The contractor is to use for the work on this stack only the best workman, skilled in such work, and he must exercise the greatest care in keeping the outside surfaces of the shaft and the inside surfaces of the core true, smooth and well pointed; and furthermore, must take great care that the outside of the stack is not disfigured by mortar, etc.

After the completion of the work, while the staging is being removed, the exterior of the stack must be well and thoroughly cleaned with acid or other proper means and then well oiled.

FOUNDATIONS.

POWER STATION.

The contractor shall provide suitable and substantial foundations for three direct-coupled 150 h. p. engines and dynamos, and for boilers.

ENGINES.—Foundations for engines to be built of hard burned brick laid in Portland cement, and to be of such dimensions as demanded by the apparatus to be selected; drawings and templates of same will be furnished by the contractor supplying the apparatus. Each foundation will be 60 inches deep unless a solid footing satisfactory to the Architect can be had with less depth; the minimum allowable depth, however, will be 48 inches. Under each foundation there will be at least 18 inches of concrete, upon which the brick work will rest. Contractor will place all anchor bolts and plates provided by engine contractor, complete ready for engine contractor to set up his machines.

Each of these foundations will contain approximately 250 cubic feet of concrete and 200 cubic feet of brick work.

BOILERS.—The boiler foundations will be built of hard burned brick laid in cement mortar, made of one part Portland cement to two parts sharp sand. The walls under the front and under the bridge walls to be 30 inches wide, and to extend to a depth of at least 9 feet below the finished floor line of the boiler room, and to be built up within 12 inches of the floor; the side and party wall foundations to be 24 inches wide, and to be tied to the rear and bridge wall foundations, but to extend to a depth of only 36 inches, same as the rear wall, provided a good material can be found at both of these depths, otherwise the contractor will increase the depth until a substantial footing is reached. Under all walls there will be at least 8 inches of well tamped Portland cement concrete which will extend at least six inches on either side of walls. The two front transverse walls to be tied together at the top by continuing the side and party walls in the form of arches, 12 inches thick at center; the intention is to have sufficient room between these two walls to run an ash car; the bottom to be tied by cement concrete at least 18 inches in thickness, extending 6 inches either side of walls. For five feet above this concrete the two 30 inch walls to be laid in Portland cement. The entire foundation to be of such dimensions as demanded by the make of boilers selected; the general dimensions, however, to be substantially as shown in the drawings.

COURT HOUSE PLANT.

HOT WATER TANKS.—For each of the two Hot Water Supply Tanks there will be provided two brick piers approximately 36 inches deep by 13 inches in thickness and 45 inches in length. Bricks to be hard-burned and laid in cement mortar; foundations to rest on 8 inches of cement concrete extending at least 3 inches either side of pier. Contractor to set all anchor bolts and plates and securely fasten the cast iron saddles in position, all furnished as elsewhere specified.

PUMPS.—For the 250 gallon electric house pump there will be built, in accordance with templates furnished, one Portland concrete foundation at least 24 inches square and 48 inches deep, 24 inches of which depth will be above the finished floor line, contractor setting all anchor bolts and plates ready for the pump contractor to place his apparatus; the corners to be rounded, and the complete foundation above the floor line to be smoothly finished with a coating of Portland cement.

The foundation for the ice water circulating pump to be the same as that for the house pump, with the exception of being at least 28 inches square instead of 24 inches.

SURGE TANK.—For the surge tank there will be built two 18 inch saddle walls 6 feet long and extended to a depth of at least 36 inches below the finished floor line. there resting on 18 inches of Portland cement concrete, extending at least 6 inches beyond piers all around; piers to extend above the floor line so as to support the bottom of the tank 30 inches above the floor, and to have a total height of 48 inches at ends. That part of piers above floor to be plastered with a finished coating of Portland cement. The tank to be supported will be approximately 5 feet diameter by 10 feet long, and the distance between the piers will be determined by the flanged openings in the bottom of tank.

AMMONIA COMPRESSOR.—The foundation for the Ammonia Compressor to be built of Portland cement concrete in accordance with the templates furnished by the contractor supplying the apparatus, who also supplies anchor bolts, plates and nuts, which the contractor will properly place complete ready for the compressor contractor to set his machinery upon the finished foundation. Foundation to be at least 56 inches deep, extending at least 8 inches above the finished floor line. In this foundation there will be approximately 200 cubic feet of concrete.

The contractor guarantees that none but the best material will be used throughout for all foundations specified under the section headed "Foundations"; the cement mortar being made of one part Portland cement to two parts sharp sand and five parts water; the brick used in all cases to be hard-burned brick; and the concrete to be made of one part cement, three parts clean sharp sand, and ten parts broken stone; the largest stone to pass through a two inch ring, and the smallest one to pass through less than a one-half inch ring; all concrete to be well tamped until cement flushes to the surface.

TUNNEL.

The intent of the following specifications is to provide for the furnishing of all·labor and material and the building complete of the Tunnel, connecting the Court House with Power Station, all as hereinafter set forth.

The contractor shall make all excavations necessary for doing the work, shall furnish all material and shall refill all excavations, remove all surplus material and rebuild the streets wherever disturbed.

ROUTE OF TUNNEL.—The tunnel shall extend from the inside wall of Court House (see point "C" on Tunnel drawing) west, crossing main sewer on Calhoun street, thence north to a point 29 feet north of the north line of Jail driveway, thence west to point "E" on the Tunnel drawing. From Court House to point where Tunnel turns north, it shall be of such construction as shown in Figure 3 and be run with level grade: the highest point of floor of Tunnel "A"

being 6 feet 3 inches below the lowest point of street grade.
Turning north the same construction shall be used to point
"D," 100 feet from center line of section of Tunnel leaving
Court House. At this point the Tunnel shall be constructed
as shown in Figure 2; this construction continuing to the
end of Tunnel in Power Station.

MATERIAL TO BE USED IN CONSTRUCTION.—
All brick to be hard burned sewer brick laid in cement mortar; cement and tests to be the same as specified for foundations for buildings; the mixture being one part cement,
three parts clean sharp sand mixed to volume; same to be
mixed dry and then wet with as little water as will make
a good working mortar.

The outside of the Tunnel is to be plastered with cement
mortar 1-2 inch thick. All mortar shall be mixed fresh for
the work in hand and any mortar which shall have set or
become hard in the box shall be thrown out and shall not be
used in this work. All bricks shall be thoroughly wet before
laying.

CONCRETE.—The 6 inch bed of the entire Tunnel and
the 3 inch layer of concrete over Tunnel from "C" to "D" to
be composed of one part cement, same as specified for foundations for buildings, three parts clean sharp sand and five
parts clipping of lime stone, granite or other hard stone,
parts mixed to volume; no stone to be larger than
such as will pass through a 2 inch ring or smaller
than will pass through a 1-2 inch ring. The cement
and sand to be mixed dry and then wet with sufficient water
to make good working mortar and then thrown on stone
and the whole turned over not less than three times. The
concrete in all cases to be thoroughly packed and tamped
with heavy rammers until water rises to the surface.

MIXING.—In the mixing of all mortar and in making
concrete, same shall be mixed in boxes or made on wooden
platform and be kept free from dirt.

"I" BEAMS.—All beams used in Tunnel construction
from "C" to "D" to be 12 inch, 32 pound "I" beams, set 24
inches from center to center; the brick arches between these
"I" beams to have a 3 inch spring.

BED OF TUNNEL.—The inside bottom of the Tunnel to be finished with a smooth cement finish 1-2 inch
thick, sloping to center line of Tunnel, so as to drain to

trough in center as shown on drawing, which trough is to drain Tunnel to sump.

MAN HOLES.—At the four points marked"MM" on Tunnel drawing man-holes shall be placed. They shall be the regulation brick sewer man-holes as shown in Figure 1, extended up to street surface and covered at top with an iron man-hole ring and a countersunk man-hole cover, (not perforated.) A special man-hole construction is provided for point marked "MM," the same being clearly shown on the drawings.

SUMP.—At north end of Tunnel at point "MM," or at some other point between point "MM" and point 'E" there shall be built a circular brick sump, 24 inches in diameter and 4 feet deep; sump to be placed in center line of Tunnel and to be suitably fitted with cast iron ring and removable grating, substantially as shown on drawing.

CABLE RACKS.—The contractor is to furnish and install on west and south walls of Tunnel iron cable racks or hook plates with capacity for six electric cables. These racks to be of such size as is used to carry 2 inch pipes with 4 1-2 centers. The lowest hook to be approximately 15 inches from floor line of Tunnel. The racks to be spaced 10 feet apart from end to end of Tunnel.

HANGERS AND SUPPORTS FOR STEAM PIPES. —The contractor will furnish and securely and properly attach to the roof of the Tunnel, where called for on the drawings, iron hangers for supporting a fourteen inch main steam pipe. There will also be furnished suitable brick piers for supporting both the fourteen inch main and the six inch return pipes, all as shown on the drawings.

Both pipes will run the entire length of the Tunnel and supporting piers or hangers will be placed ten feet apart. The exact location as to height is to be as demanded by the steam heating specifications, the grades being given in such specification.

The hangers and roller supports to be made in accordance with the drawings, but in case the size of the fourteen inch pipe is reduced, the design of hangers will be modified to suit the change, but will be substantially that as shown and of the same length, in either case to permit of a twenty inch lineal expansion for either pipe. Roller bearing plates to be at least fourteen inches long for the first seven hundred feet from Court House, and at least twenty-four inches long for the remaining distance. Where cast iron plates are shown in that part of the Tunnel from point marked "P" to the Power Station same will be laid in the concrete bottom as shown.

All saddles and roller plates together with sliding plates are to be provided with planing strips, to be truly and acurately planed. All rollers to be turned and finished accurately to a diameter of one and one-quarter inch.

PERMITS.—The contractor shall obtain the necessary permits from the City authorities for the performance of his work.

ITEMS NOT INCLUDED.

It is intended that the foregoing specifications, together with the drawings, shall cover all points of construction and finish throughout the entire work, with the exception of the items noted to the contrary in said specifications, and the following: Decoration, Electric Illuminating Fixtures and switches, Office Counters and Screens, Screen Partitions, Cashiers' Cages, Vault Doors and Lining, Spiral Stairs to Vault, Bar Railings, Judges' Benches and Court Room Furniture, Sidewalks and Curbing, and Glass and Hardware as specified under their respective headings.

ADDENDA

CORLISS ENGINES, BELTING, DYNAMOS,

ENGINE AND DYNAMO FOUNDATIONS,

. . . AND . . .

ENLARGING OF POWER STATION.

The specifications which follow as addenta to the preceeding specifications, are intended to govern propositions for the furnishing of a complete Corliss Engine plant, and necessarily cover the required belting, additional Engine and Dynamo Foundations, together with such changes as would be demanded in the Dynamos by the adoption of such a plant, and the enlargement of the Power Station.

ENGINES.

GENERAL.—Propositions are invited upon Corliss Engines as hereinafter set forth, to be set up complete upon foundations to be built in accordance with templates furnished by the contractor. The engines shall be complete in every detail, and turned over to the County in such condition ready for operation.

Contractor shall furnish an expert to erect, adjust, start and run the engines for one week, or until such time thereafter as may be necessary to put each engine in successful working condition. The expert to instruct the County's engineer as to the proper method of operating engines, or any other apparatus which he may furnish under these specifications.

NUMBER.—The number of engines to be furnished under these specifications is three.

TYPE.—Engines to be simple, slow speed Corliss, of the latest and most approved pattern, especially adapted for electric lighting service, all three engines to be right handed.

RATING.—Each engine shall develop one hundred and thirty-three indicated horse power when cutting off at one-fifth stroke with ninety pounds initial steam pressure, with engine running at a speed of seventy-eight revolutions per minute.

FITTINGS.—Each engine shall be provided with the following: One throttle valve fitted to the steam chest. One exhaust valve (straightway). One automatic sight-feed cylinder lubricator. Full set of graduating oil cups. One set of wrenches. Drip collectors of polished brass. Polished brass splasher. One full set of brass hand oilers with tray.

All necessary pipes, valves and attachments permanently connected with the engine, for taking indicator cards from both ends of the cylinder; these connections to be fitted with a three way cock. All this piping including fittings, to be nickel-plated. In addition to the above each engine to be provided with nickel-plated standard "Crosby" water relief valves, which may be attached to indicator piping if so desired by Engine builder, and fitted to ends of cylinder.

Contractor shall also furnish and attach all requisite drain valves and pipes, fitted to steam chest and ends of cylinder.

With each engine as provided, there will be furnished and attached an approved automatic stop to prevent the running away of engine.

Contractor shall furnish one perfect reducing motion in the form of an approved pantograph provided with necessary stands arranged for each engine, and all requisite attachments for connection to cross-head.

With the foundation templates to be furnished, there will be provided all necessary foundation bolts and plates for all engines.

FLY-WHEEL.—The fly wheel of each engine to be

twelve feet in diameter, with not less than a twenty-three inch face.

The fly-wheel is to be used as a driving pulley, and shall have ample strength and weight, be true, faced and shall be well chowned.

FINALLY.—The engine shall operate noiselessly and without pounding or vibration. All parts shall be accurately made to standard guage and interchangeable, all moving parts carefully balanced, and all valves and packing free from leakage.

Engines shall be capable of being run continuously under full rated load for twenty-four hours without undue heating of journals or bearings. The engine foundations to be furnished as elsewhere specified, the contractor furnishing complete working drawings for same.

The contractor shall furnish such renewal parts as he deems advisable to keep on hand, specifying each item separately.

GUARANTEES.—The contractor shall guarantee for each engine the following:

1. The steam consumption per indicated horse power per hour, when operating under normal and regular working conditions with initial steam pressure of ninety pounds at one-fifth cut-off and full rated load, steam not to contain more than one per cent. moisture.

2. Same as guarantee 1 when engine is operating at fifty per cent. of rating as called for.

3. The per cent. variation in speed of engine through any changes of load from no load to full load, steam pressure varying from a minimum of eighty-five pounds to a maximum of ninety-five pounds.

4. That material, workmanship and finish shall be the best throughout.

5. To hold himself responsible for faulty workmanship, and to replace and make good any defects which may develop within one year from date of final acceptance of his apparatus, without cost to the County.

BELTING.

The contractor shall furnish and deliver in position, ready for service, three endless double-leather belts for driving dynamos direct from fly-wheels of engines.

Belting shall be made of pure oak bark taned leather of

positive center cuts, must be strictly short lap, no piece to
exceed fifty-four inches in length, including the laps, and
shall be made free form any shimming or filling whatever;
shall be free from rivets, pegs, sewing or fastening of any
kind save that of cement; and must be capable of trans-
mitting the required horse power of engines with an ample
factor of safety.

The belts are to be of such lengths as required by the position of machinery. The bidder shall state the weight per
square foot of the belts he proposes to furnish and give the
price per foot at which he will furnish the belts as called
for. Belts will be twenty-two inches in width, with
twenty-two foot centers, driving from a twelve foot fly-
wheel to a twenty-one inch dynamo pulley. The bidder will also state the price per foot of twelve inch belts,
as the selection of dynamos will determine whether one or
two machines will be driven from each engine. All belts
in any case to be made endless and put in position at station,
by contractor.

The contractor shall guarantee his belts to be evenly
balanced throughout; to be of uniform thickness; to be
free from brands or cuts; and to conform to the requirements of these specifications, and to hold himself responsible for faulty workmanship or stock and to replace the
same within one year from the date of final acceptance,
without cost to the County.

DYNAMOS.

In the event of a Corliss Engine Plant being installed,
the dynamos shall conform in all respects to the Specifications for Dynamos, with the following exceptions:

If Corliss Engines are installed, the dynamos will be belt
driven instead of direct coupled, and shall be designed to
run at a speed of not more than four hundred revolutions per
minute. Each dynamo to be equipped with suitable insulating base frame and belt tightning device, and to be provided with a suitable cast iron pulley of twenty-three inches
face, and of proper diameter to operate dynamo at its specified voltage when driven by belt from twelve foot driving
wheel having a speed of seventy-eight revolutions per minute.

If Corliss Engines are installed, and the electrical system adopted shall be a three wire system, with two dyna-

mos for each engine, then the speed of the dynamos shall be not more than four hundred revolutions per minute, and shall be as above specified, excepting that each dynamo shall have one-half the ampere capacity called for in the Dynamo Specifications, and shall be provided with a pulley having a thirteen inch face.

FOUNDATIONS.

ENGINE.—The foundations required for each engine will contain approximately ten thousand and forty cubic feet of concrete, and twenty-three thousand four hundred brick. The concrete and brick work to conform to the specifications providing for foundations for direct-coupled sets. Each foundation will be provided with a cut cap stone at least eight inches in thickness.

DYNAMO.—Each dynamo foundation will require approximately sixty cubic feet of concrete and one hundred and seventy cubic feet of brick work. The concrete and brick work to conform with the foundation specification relating to direct-coupled sets.

POWER STATION.

In case Corliss Engines should be adopted, it will necessitate the enlargement of the Machinery Hall of the Power Station from the dimensions shown and figured on drawing No. 69.

Bidders are requested to submit proposals covering all necessary and required work and material in making such change, and must include additional excavation and concrete under the smoke stack on account of the descent in grade at that point caused by the extension farther west than shown on the present drawing. Also, additional excavation, concrete footings and foundation work must be considered as well as all required additional drains, brick and brick work, enameled brick, lime stone and cutting, architectural terra cotta, structural and ornamental iron work, slate roofing, copper work, sheathing and felt, mosaic tile floor, metal ceiling, marble work, carpenter, joiner and cabinet work, and all other work and material required in the work as already specified and shown on the drawings, together with any other work and material that will be requisite and necessary in making the change.

The increased dimensions will be in the Machinery Hall, which will be fifty-eight feet and six inches wide instead of fifty feet and six inches wide as now shown, an increase in width of eight feet, and an increase in depth from east to west of thirty-four feet. These are inside measurements between the enameled brick lining.

The heights will remain the same excepting the foundation walls where it becomes necessary to conform to grades. and the roof, which must retain the same pitch as now shown, and will consequently be higher at the appex. The roof must be calculated for a load of fifty pounds per square foot, exclusive of the weight of the structural material entering into its use, and the metal ceiling suspended therefrom.

The roof must be entirely supported on the four walls forming the Machinery Hall, and must not have any intermediate or column supports, as this room must be left entirely free of all obstructions.

All detail features of this room to remain as now shown, including balcony, screen partition, stairs. etc.

Provision must also be made in the bids for two additional windows on the north side and two additional windows on the south side, making four windows and one door on each of these sides instead of two windows and one door on each of these sides as now shown. Allowing also for additional pilasters, etc., between windows.

Bidders are required to fill out the schedule for this work, which will be found attached to the general schedule.

The boiler room, office rooms and entrance vestibule to remain as shown on plans, excepting the boiler room where changes in the foundations will be necessary as specified.

INDEX.

INDEX TO ADDENDA.

SUPPLEMENT

TO

SPECIFICATIONS

FOR A

NEW COURT HOUSE,

HEATING, LIGHTING AND POWER PLANT

AND

. . TUNNEL . .

FOR

ALLEN COUNTY,

INDIANA.

1897.
THE JOURNAL CO., PRINTERS,
FORT WAYNE, IND.

Instructions and Conditions.

The following specifications are to be considered part of the general drawings and specifications approved by the Board of Commissioners of Allen County, Indiana, January 9th, 1897, and now on file in the Auditor's office of said county.

Said drawings and specifications will remain in full force and effect except drawing No. 69, and such changes that are specifically mentioned in the following specifications.

No other drawings have been prepared, except No. 71 which takes the place of No. 69, which is omitted, and the Power Station will therefore be erected according to drawing No. 71 and these specifications.

All proposals must be based upon the requirements of the drawings and specifications, and should any changes be made after the work is contracted for, deductions and additions in price will be made, as set forth in the General Instructions and Conditions.

The contractor shall furnish and provide all necessary and required work shops, etc., on the Court House premises, for modeling, etc., and scaffolding for the execution of all interior and exterior work.

The addenda specifications will remain in full force and effect, with the exception of that portion referring to the Power Station, which will be omitted.

The time of completion of the Court House is extended to the first day of November, 1899.

EXCAVATION AND FILLING.

The excavations for Court House foundations will be three feet less in depth than figured on drawings. The height from Basement Floor to Ground Floor will be seven feet and eight inches, and the bottom of all footings three feet and ten inches below Basement Floor level.

The contractor may have free use of all surplus soil from

Court House excavations, and the brick spawls from old court house for filling and grading in and about the Power Station and Stack. The brick spawls are now near the site of the proposed Power Station and Stack.

DOWN SPOUTS AND DRAIN TILE.

The down spouts of Court House will not be connected with the drain tile as before specified, but connected as specified in Plumbing Work. Neither will the joints of drain tile be cemented, but the tile laid close together and bedded in coarse sand and gravel.

FOUNDATIONS.

The Power Station foundation will be of hard blue lime stone instead of concrete and brick, as before specified. The footings to be large, flat stone, each stone filling the course in width and height, close fitted and flushed up with spawls and mortar, laid on natural undisturbed earth. The walls, up to the water table, to be of good, large, sound building stone, well bonded their full length every three feet in height. All stone to be laid on their natural beds to a line on both sides, the joints well filled with mortar and well trowel pointed on both sides. All mortar to be composed of one part Louisville or Milwaukee cement and three parts clean, sharp sand. The outside surface of walls will not be plastered as before specified.

The foundations for boilers and stack to remain as called for in the original specifications. All cement used in foundation for Stack to the top of stone base course at grade level, and the foundations for boilers to be best American Portland cement. The boiler foundations and ash tunnel to be scheduled with the Power Station foundations.

The Court House foundations will be three feet less in depth than figured on drawings. The height from Basement Floor to Ground Floor will be seven feet and eight inches, and the bottom of all footings three feet and ten inches below Basement Floor level.

All cement used in footings to be German Portland cement, of either the Josson, Dyckerhoff, Alsen, or Germania brands, and at all times must show neat 400 lbs. per square inch tensile strength. All footings and cast iron shoes above Basement Floor level to be thoroughly protected with neat cement.

All cement used in outside walls and piers between top of footings and top of granite base course, and in all interior piers, except the eight central tower piers, between top of footings and Ground Floor level, to be best American Portland cement. The outside walls and piers to be plastered as before specified, with the same kind of cement. The retaining walls between piers and under entrance steps may be sixteen inches thick instead of the thickness figured on drawings. Brick arches will be sprung between piers to support the work above. To conform to the jointing of steps, intermediate walls will be constructed under them.

All cement used in the eight central tower piers between top of footings and Ground Floor level to be German Portland cement, as above specified for footings. All brick used in these tower piers to be best quality selected hard sewer brick.

Ninety per cent. by weight of all cement shall pass a No. 100 sieve, 10,000 meshes per square inch.

All brick work to be thoroughly grouted with mortar as specified.

BRICK WORK ABOVE FOUNDATIONS.

The shaft of Stack will be octagon on the exterior, instead of circular, as shown on drawings, the interior to remain circular, as shown. Special brick to be used for the octagon work.

All brick work in Stack, except fire brick, above top of stone base course at grade level to four feet above stone base course of shaft to be built up with mortar composed of one part Louisville or Milwaukee cement and three parts lime mortar, the lime mortar to be composed of best quality fresh burned lime and clean, sharp sand.

All brick work above four feet above stone base course of shaft, except fire brick, and above stone foundations of Power Station, to be built up with best quality lime mortar. The face brick to be selected hard common brick of uniform color. There will be no interior lining of enameled brick, as before specified.

All brick used in Power Station and Stack, except face brick and fire brick, may be those taken from the old Court House, of which there is supposed to be a sufficient quantity acceptable for this work. Should there not be, the Commissioners will allow the contractor the difference in price for new brick. The Commissioners will dispose of these old brick to the contractor for $3.00 per thousand. They are now near the site of the proposed Power Station and Stack.

All brick work in Court House above foundations, except the eight central piers of tower to roof lines, to be built up with best quality lime mortar, the mortar to be composed of best quality fresh burned lime and clean, sharp sand.

All brick used in Court House above foundations, except the eight central piers of tower to Court Room Floor level, to remain as called for in the original specifications.

All brick used in the eight central piers of tower to Court Room Floor level to be best quality selected hard sewer brick laid in German Portland cement, as specified for footings and foundations of these piers. From Court Room Floor level to roof lines these piers will be built up in best quality American Portland cement.

Ninety per cent. by weight of all cement shall pass a No. 100 sieve, 10,000 meshes per square inch.

All brick work to be thoroughly grouted with mortar as specified.

There are no large brick arches in the tower, as originally specified.

GRANITE.

The base course, steps, platforms, plinths of Court Street entrance, and all of the Calhoun, Main and Berry Street entrance work to be granite, as before specified.

The Court Street entrance work above the granite plinths to be Blue Bedford stone, instead of granite, as before specified.

Only the granite columns of the Calhoun, Main and Berry Street entrances are detached, as shown on drawings. The granite plinths at the Court Street entrance to be for three-quarter diameter columns attached to the pilasters.

The jambs of all these entrances to be bed jointed with the Basement Story stone work and properly bonded to the same.

The sides of pilaster caps will not be carved, except the mouldings, as shown.

BLUE BEDFORD STONE.

The Court Street entrance work above the granite plinths to be Blue Bedford stone, instead of granite, as before speci-

fied. The columns at this entrance to be three-quarter diameter cut on the pilasters. The jambs of this entrance to be
bed jointed with the Basement Story stone work and properly
bonded to the same. The sides of pilaster caps will not be
carved, except the mouldings, as shown. All this work to
be rubbed, tooled and carved the same as specified for granite work, except the columns, which will be tooled ten cut
vertically, instead of polished.

The granite base course and plinths and the moulded base
course and bases of columns, imposts, etc., on the granite
base course and plinths to be bedded and pointed in mortar
composed of Lafarge cement one part, and clean, sharp
sand three parts, and the backs plastered a thick coat of
same mortar.

No Lafarge cement to be used above the moulded base
course, but to be bedded and pointed in lime putty mortar
made of fresh burned lime and clean, sharp and white lake
sand or marble dust, screened through a fine mesh sieve.
The back joints to be thoroughly slushed with mortar.

All vertical joints in ashler to be as directed. The courses
between Basement Story windows to have alternately one
and two vertical joints. The balance of the work to be
jointed correspondingly. Plumb bond to be maintained
throughout the entire building where practical.

The large arches over entrances will remain as shown on
elevations, instead of lowered, as before specified and shown
on drawing No. 61. There will be no frieze or inscription
between the arches and cornice. The additional heights of
columns, entablature, etc., will be maintained, as before
specified and figured on the various drawings.

Plaster models of mouldings will not be required, except
where they come in connection with sculpture and carving.

The contractor will be required to do all stone sawing
within the corporate limits of the city of Fort, Wayne, Ind.
He will also be required to employ on the cut stone work
none but first-class stone cutters.

The stone work for the Power Station will consist only of
the water table, steps and door and window sills, as shown
on drawing No. 71. The stone work for Stack to remain as
shown on drawing No. 70.

SCULPTURE AND CARVING.

The large arches over entrances will remain as shown on

elevations, instead of lowered, as before specified and shown on drawing No. 61.

The spandrels of Court Street arch to be sculptured, instead of carved, as indicated on elevation.

The sculpture in panels of balustrade to occupy all the space between the carved mouldings forming the panels and not only the plain raised surfaces, as shown on drawing No. 63.

The sculpture in pediments to have not less than fifteen inches relief, and the sculpture in balustrade not less than eight inches in relief. All other sculpture to be in proportionate relief.

All sculpture to be executed after the material for same is secured in its permanent position and all other exterior work is sufficiently advanced to permit of its execution.

Separate models must be made for the sculpture in each and every pediment, panel in balustrade, spandrels of arches and all other sculpture, as no duplicates will be accepted. The busts must be portraits of different personages.

Models of all important pieces of carving will be required.

All modeling to be done on the Court House premises and placed in proper height and distance position for inspection.

Bidders are required to mention in their proposals the names of the modelers, sculptors and carvers they purpose doing the work, but the acceptance of any proposal by the Commissioners must not be considered by the contractor as an approval of any modeler, sculptor or carver, as none but thoroughly competent and high class artists will be approved by either the Commissioners or Architect.

ARCHITECTURAL TERRA COTTA

There will be no architectural terra cotta in the Power Station. That in the Court House will remain as already specified and shown on the various drawings.

STRUCTURAL IRON WORK.

As the Court House foundations will be three feet less in depth than figured on drawings, all cast iron columns in Basement Story will be three feet less in height and the stairs leading from Ground Floor to Basement Floor two feet less.

Instead of four flag staffs, as originally called for, there will be only one, and this on the Calhoun street front.

The gas pipe guard rails around all ceiling lights in attic will be omitted, except the guard rail on balustrade around ceiling light or inner dome of rotunda, which will remain as originally specified and shown on drawings.

The wire netting that was originally specified for this balustrade and all other balustrades over inner dome will be omitted.

All stairs, platforms, balconys and floors around and over inner dome to have raised skirtings to protect the glass below from dirt.

The balusters of all balustrades around and over inner dome to be not more than six inches between centers, instead of twelve inches, as before specified.

The awning fasteners for all east, south and west windows and the metal bars for glass shown on drawings No. 11 and 12 will be included in the carpenter and joiner work, and all metal lath will be included in the plastering.

Cast iron and gas pipe railings for Power Station will be as shown on drawing No. 71.

Wherever nut locks are required, the Buel nut lock is recommended.

Leave no obstructions to Mail Chute.

ORNAMENTAL IRON WORK.

The balustrades of stairs from Sheriff's Office to Court Room Floor, in Auditor's Office to Ground Floor, from Court Rooms Two and Three to Jury Rooms in attic, the spiral stairs and the balustrade around balcony in Law Library to be cast iron, instead of wrought iron, as originally specified. All of these stairs will have marble treads, platforms, risors, and wall bases, except the spiral stairs, which will have marble treads only.

All of the above mentioned work and that forming the inner dome, cornice under same, crown mouldings of balcony cornice under inner dome, the stair and floor facias connected with the above mentioned stairs and balustrade and the mouldings around all ceiling lights, excepting in the four court rooms, will not be bronze plated, as originally specified, but finished in artistic cast iron work and painted one coat of mineral paint before erection.

All of the other work, material and finish, to remain as

called for in the original specifications, with the following exceptions: The balustrades under inner dome, over grand stair case and around lantern of dome, the panels of marble balustrades on Office and Court Room Floors in rotunda to be cast iron, of fine detail, instead of wrought iron, as originally specified. The balustrade around lantern of dome and the crown mouldings of the cornice of dome to be duplex bronze plated and the private elevator enclosure Bower-Barffed. The entrance doors to be made of galvano-bronze with special bronze hardware, trims and sashes of iron duplex bronze plated. The grills in doors and transoms will be omitted.

All of the ornamental iron work and finish must be from one shop, as it must be specially prepared in detail, construction and finish.

Bidders are required to mention in their proposals the names of the sub-contractors they purpose doing this work, but the acceptance of any proposal by the Commissioners must not be considered by the contractor as an approval of any sub-contractor, as none but a thoroughly equipped, competent and reputable concern will be approved by either the Commissioners or Architect.

The metal bars for glass shown on drawings No. 11 and No. 12 will not be included with this work.

There will be no ornamental iron in the Power Station.

CONSTRUCTIONAL TERRA COTTA.

The exhaust fan chambers in attic to be included with this work.

Omit the tile partitions in Basement Story shown on drawing No. 32.

Construct openings in Office and Court Room Floors for Mail Chute, and cut and finish an opening in each floor directly in front of and centered upon this surface; these openings will be neatly and easily finished and their size 9x3 5-8, and shape determined by setting in them thimble of iron, which will be furnished and delivered by the contractors for Mail Chute.

The cast iron columns in Basement Story and the girders in Basement Story ceiling need not be covered with tile, but the flanges of beams between floor arches will be covered.

All porous terra cotta furring of exterior walls on the in-

side of rooms and closets, as originally specified but not
shown on drawings, will be omitted.
. Excepting those in the ceilings of Vestibules, Lobbies,
Rotundas, Corridors, the Four Court Rooms and the Commissioners' Court Room, ceiling of Grand Staircase Hall
and the upper ceiling of the Private Staircase Hall, all false
beams in the ceilings of Ground, Office and Court Room
Storys will be omitted. This includes the false beams forming panels in the offices of the Sheriff, County Superintendent, Township Trustees, County Assessors and the Assembly
Room on Ground Floor, the Auditor, Treasurer, Clerk and
Recorder on Office Floor, and the Law Library on Court
Room Floor.

Excepting in Vestibules, Lobbies, Rotundas, Corridors,
the Four Court Rooms and the Commissioners' Court
Room, all furring in the Ground, Office and Court Room
Stories, forming false piers, pilasters, and one false column in
Treasurer's Office will be omitted, excepting that which is
necessary to form door and window openings, and enclosing
down spouts, ventilating pipes and flues.

TILE AND SLATE ROOFING.

The four deck roofs of the Court House will be covered
with plates of terra cotta roofing tile, size 6x9x3-4 or,
6x12x1 1-4, instead of slate, as called for in the original
specifications. In every other particular these roofs to conform to said specifications.

·The four main, or pitched, roofs of the Court House will be
covered with square weather end terra cotta shingle tile, size
6x12x3-8 or, 6x13x3-8, laid four and one-half inches to the
weather. Each tile to have two countersunk holes and securely nailed, not bolted, to porous terra cotta book tile with
two copper nails with heads to fit countersunk. The copper
nails to be of approved size and sufficiently long to extend
through the shingle tile, felting, concrete and into the book
tile sufficiently to thoroughly secure the work. Wherever
the iron work of roof, or other obstructions may interfere
with the use of nails, strong copper wire must be used and
the tile securely tied from the underside of roof. Over the
concrete on book tile there will be laid a covering of best
heavy quality asbestos roofing felt, with sufficient lap, securely fastened to the roof before placing the shingle tile in
position. The tile at eaves, skylights and walls to be bedded

and pointed up in approved elastic roofing cement in a manner which will thoroughly exclude the entrance of water by back flow or otherwise.

All of the above mentioned tile to be of a natural rich dark red, uniform in color, size, shape and texture, to be fine moulded and of great strength and durability. The surfaces to be smooth and out of wind, edges and tails straight and square, and the corners full. All corners to be perfectly straight and horizontal and the bond perfect.

All tile to be made of fine clay burned to a vitrified, close, strong body, thoroughly hard and practically non-absorbent. All to be equal to the best quality of Akron, Ohio, vitrified roofing tile.

The roof of Power Station will remain as called for in the original specifications, except that the slate may be either Peach Bottom, Monson, Main, Star Quarry, Bangor, Pa., or Buckingham, Va., and the hips and ridges will be galvanized iron, instead of terra cotta, and the flashings, valleys and gutters tin, instead of copper.

The work on all roofs to be thoroughly well done to prevent leakage.

By mistake a tar and gravel roof is noted on drawings Nos. 49 and 50.

METAL WORK.

The copper gutters of Court House to run up sufficiently under the shingle tile to prevent leakage from overflow, and extend under said tile at tower, pediments and skylights twelve inches. There will be only one flag staff, instead of four. All skylights will be copper. By mistake some are noted on the drawings to be galvanized iron.

The copper work of Stack to remain as called for in the original specifications.

All gutters, valleys and flashings of Power Station will be tin, and the hip and ridge rolls and gutter screens will be galvanized iron instead of the material originally specified.

There will be no metal ceilings in the Power Station.

All gutters, valleys and flashings of Power Station to be best quality American IX tern plates, no wasters, the brand and maker's name stamped on each sheet. N. & G. Taylor's Old Style or Gilpertson's Old Method will be acceptable. The gutter lining to extend up under the slate not less than eight inches, be properly fitted to the gutter bed and secured

to the edge of galvanized iron cornice with double locked
joint malleted close. Paint all tin gutters on the under side
and on the top where it will be covered and all tin flashing
and valleys both sides with a coat of best metalic paint and
linseed oil.

The galvanized iron cornice to be properly secured in
place, all joints to be solid, and rivited and soldered as required. All galvanized iron cornices, hip and ridge rolls,
and aprons to be No. 24 B. W. G.

CONCRETE FILLING.

The concrete filling will be as called for in the original
specification, only that it will apply to all cement and marble
floors, as well as mosaic tile floors. All cement used in this
work will be best American Portland cement.

CEMENT FLOORS AND BASES.

The entire floor areas of the Power Station and Court
House Basement will be cement; also, the Sheriff's closet,
Clerk's Storage Room, Janitor and Engineer's Room and
closet, including clock room, County Assessor's Room and
closet, Auditor's Storage Room and closet off passage way,
Township Assesor's closet, County Superintendent's closet,
Coroner's Room and closet, Public and Private Lavatorys,
Janitor's closet, Wowen's Private Lavatory, Women's Lav-
atory, Assembly Room and closet, and Township Trustee's
closet, all on Ground Floor. On the Office Floor, the Clerk's
Work Room closet, Clerk's Office closet, Auditor's Work
Room closet, Treasurer's Office closet. On the Court Room
Floor, the Jury Room closet, Judge's Chamber closet, Sher-
iff's coat room, Detained Witness' Room closet, Judge's
Chamber closet, Law Library closet, Judge's Chamber
closet, Witness' Room closet, Judge's Chamber closet and
coat room, Jury Room closet, Witness' Room closet, Witness' Room closet, Attorney's Ante Room closet, Janitor's
closet, Grand Jury Room closet, Witness Room closet and
Witness Room closet. On the Attic Floor, the Jury Room
No. 2 closet, Jury Room No. 3 closet, Gallery and Alcoves
of Law Library, Balcony over Grand Staircase, Attic Room
between stairs from Court Room Floor and stairs to Tower,
Balcony under inner dome and the floor around inner dome.

All of the above mentioned floors to be as originally specified for the Basement Floor, except that the finishing coat

may be one inch thick instead of two inches, and the cement either Louisville or Milwaukee cement.

All of the above mentioned floors, except in Lavatorys, to have cement bases, or skirting, six inches high, as originally specified for Basement Floor. Most of this work was originally specified to be mosaic tile and marble bases.

The cement used in concrete bed for these floors will be best Amrican Portland cement, otherwise as originally specified.

MOSAIC TILE FLOORS.

The mosaic tile floors will consist of the entire floor areas of Vestibules, Lobbys and Rotunda on Ground Floor, the Lobbys, Rotunda and Commissioners' Court Room on Office Floor, and the Rotunda and Corridors on Court Room Floor. All of these floors, concrete bed and other work and material connected therewith to remain as called for in the original drawings and specifications.

The other floors originally specified to be mosaic tile have herein been changed to cement and marble.

Properly trim around mail chute on Office and Court Room Floors, and cut and finish an opening in each floor, directly in front of and centered upon this surface; these openings will be neatly and easily finished and their size, 9x3 5-8, and shape determined by setting in them thimbles of iron, which will be furnished and delivered as desired by the contractors for Mail Chutes.

The mosaic tile to extend under the doors.

PLAIN MARBLE WORK.

The marble floors will consist of the entire floor areas of Sheriff's Office and Passage Way, Township Assessor's Office and Passage Way, County Superintendent's Office and Township Trustee's Office on Ground Floor. The Clerk's Document Room, Work Room, Passage Way and Office, Bridge Superintendent's Office and Private Office, Passage Ways, Lavatory, Commissioners' and Auditor's Private Office, Auditor's Office and Work Room, Surveyor's Office, Recorder's Office, Private Office and Closet, Men's Private Lavatory and Treasurer's Office, Work Room and Private Office on Office Floor. The Jury Room and Ante Room, Judge's Chamber and Alcove, Sheriff's Room; Detained

Witnesses' Room, Judge's Chamber, Law Library and Alcoves, Judge's Chamber, Stenographer, Witness Room, Judge's Chamber and Alcove, Jury Room and Ante Room, Consultation Room, Witness Room, Witness Room, Consultation Room, Witness Room, Attorney's Room and Ante Room, Grand Jury Room, Prosecuting Attorney's Room, Witness Room, Witness Room, Consultation Room and Consultation Rooms, Jury Retiring Rooms and Closets in Court Rooms Nos. 2 and 3, and the four Court Rooms and Alcoves on Court Room Floor, and the Jury Rooms and Vestibules on Attic Floor.

All of these floors to be tiled with best quality dark clouded Vermont or gray Tennessee marble tile twelve inches square and seven-eighths inch thick, joints in true line lengthwise and to alternate crosswise. Borders to be long lengths, of equal widths on opposite sides and of such widths that only exact whole and half tiles may be used against borders.

The portion of borders between door jambs to be in one piece and extend under the doors.

There will be dark clouded Vermont or gray Tennessee marble thresholds, six inches above the floor, to all doors entering closets and stalls, in which water closets are placed on Office and Court Room Floors. These thresholds to be the full width of doors and extend three-quarters of an inch beyond the face of platform risor, the front edge to be rounded. The back edge to be jointed with the cement and marble floors of closets and stalls and on a line with the cement and marble bases. These thresholds to be one and one-quarter inch thick, strong.

All tiles to be true and out of wind, edges square, backs sawed fair, joints to be close, beds full, the surfaces of all joints smooth and even, and after laying, all joints to be grouted full with thin cement mortar.

The cement used in concrete bed for these floors will be best American Portland cement, otherwise as originally specified for mosaic tile flooring.

Including the water closet platform risors, the wall bases on the above mentioned marble floors, except those specified below, to be pink Tennessee or medium clouded Vermont marble. The bases to be eight inches high and the top corner rounded.

The bases in Court Rooms Nos. One, Two, Three and Four, and in the Commissioners' Court Room will be dark

Tennessee or Swanton Black Vermont marble. All to be of the dimensions shown on drawings. The base in Commissioners' Court Room to have double plinth, as shown. The floor bases of all columns will be marble, corresponding with the wall bases opposite.

The stairs from Sheriff's Office to Court Room Floor, in Auditor's Office to Ground Floor, from Court Rooms Nos. Two and Three to Jury Rooms on Attic Floor and the two spiral stairs in Law Library will have marble treads, platforms, risors and wall bases, except the spiral stairs, which will have marble treads only. The treads and platforms to be medium clouded Vermont or gray Tennessee marble one and one-half inches thick, front edges rounded. The risors and wall bases to be pink Tennessee or medium clouded Vermont marble.

All windows throughout the Court House will have marble stools and aprons, except windows at sides of interior doors, to be blue Vermont or pink Tennessee marble, except in Court Rooms Nos. One and Four, and the Commissioners' Court Room, where they will be white Vermont or white Italian marble. The jambs, soffits and reveals of outside windows in Commissioners' Court Room and the plain surfaces between bases and aprons under said windows, and the jambs, soffits and reveals of doors and windows in Court Rooms Nos. One, Two, Three and Four will be scagliola, instead of marble, as shown on drawings.

The Lavatories on Ground Floor, the Lavatory on Office Floor, the closet in Lavatory off Commissioners' Court Room and the closets in Jurors' Retiring Rooms off Court Rooms Nos. Two and Three, to be wainscoted with marble seven feet high on all sides, to have caps and bases, bases ten inches high, caps six inches high, both to project one-quarter inch from face of wainscoting with corners rounded, wainscoting to be flush with plaster surface above. The bases in Lavatory near said Commissioners' Court Room and Jurors' Retiring Rooms, to be ten inches high, top corners rounded. In the above mentioned closets the partitions will be marble of the same thickness, height, finish and design of door opening, as marble fronts of water closet stalls figured and shown on drawing No. 31. The dimensions are to be as figured on this drawing and not as originally specified. Marble stalls for urinals shall be five feet eight inches high, partitions two feet centers, and two feet in width, the middle partitions to be ten inches from the floor and the end partitions on the floor. Partitions to be seven-eights inch thick, marble bases to be two feet two inches wide, and the length required, and two inches thick, countersunk one-half inch. All the marble mentioned in this paragraph to be best quality light pink or gray Tennessee or light clouded Vermont marble. All to be furnished ready for setting and set by the marble contractor, except the above mentioned partitions in Commissioners' and Jurors' Lavatorys, the water closet stalls, partitions, fronts, backs and ends, and the urinal partitions, ends, backs and bases, which will be set by the plumbing contractor. The basin slabs, backs, ends and aprons to be of the same kind of marble as the other work, furnished and set by the plumbing contractor. The mirrors and minor frames and the drinking fountains are omitted entirely from this work.

All marble to be the best quality, and samples to be submitted with proposals.

There will be no marble in Power Station, except one wash basin, furnished and set by plumber.

ARCHITECTURAL MARBLE WORK.

On Court Room Floor between marble sub-plinth course and plaster entablature on Court Room sides of Rotunda between marble tower piers and on Law Library side of Rotunda between marble pilasters opposite marble tower piers, all work therein originally specified to be marble, is now changed to scagliola and plaster. This consists of the work around doors entering Court Rooms No. Two and Three and Law Library, fluted pilasters at each side and tablets over the doors, window and tablet architraves, etc., and the plain wall surfaces.

On drawing No. 17, the frieze of entablature is noted in one place to be marble. It should be plaster.

The caps above astragals of pilasters at Ground Floor starting point of East Vestibule stairs, the brackets over these caps and those opposite at intersections of beams between these brackets and the beams spanning between tower and arch under said stairs and the soffits of these beams to be plaster, instead of marble, as originally intended and shown on drawings Nos. 18 and 23. The facias of these beams, the brackets on said arch and the brackets under landings between Office Floor and Court Room Floor to be -

marble, as originally intended and shown on said drawings. The Dolphins on stair rails and newels under small marble columns will be omitted.

The spandrel panels formed by the intersection of ceiling plaster beams, stair soffit marble beams and ceiling plaster beams, and the circular balustrades in both of North and South Lobbies and East Vestibule stairs shown on ceiling plans of stairs on drawings Nos. 22 and 23, will be plaster, and not marble, as one of them is noted.

The Greek fret on facia under arch window over East Vestibule stairs will be omitted. The window will be raised about two feet, as specified in stone work. This will allow the marble plinth over facia to continue under the arch window without forming a window sill or stool. The scagliola base of pilasters will not continue. There will be a marble stool and apron under the raised window to stop or return on themselves between the scagliola pilasters on each side of the window. The additional surface between plinth on facia and the apron under window stool will be scagliola formed in panels.

Omit all carving on mouldings, egg and dart, forming panels in tower piers, as shown on drawing No. 17. The moulding to remain in profile.

The caps of wide pilasters forming jambs of entrances to Judiciary Corridors will be marble, instead of plaster, as noted on drawing. Omit enrichment of mouldings. The soffits over these will be marble, except panels, which will be plaster.

The two members of small ornamented moulding between plaster panels and architraves over entrances to Court Room Floor corridors will be scagliola, instead of marble, as indicated by witness lines on drawings Nos. 17 and 18.

Omit the carving on mouldings of architraves over elevator fronts on Court Room Floor.

Omit the carving on starting newels of all stairs on Ground Floor.

Omit the beaded mould immmediately over architraves of doors on each side of East Vestibule between Court Street entrance doors and stair arch, Ground Floor.

Omit the beading of moulded architrave around entrance door to Commissioners' Court Room, Office Floor.

The architraves of doors, windows and elevator fronts on Ground Floor, instead of being moulded, as shown on drawings, will be made up of five plain facias, including one formed by the thickness of jambs and soffit and window sill piece, to increase in width from the inside to total width shown. The window architraves to continue on the under side, omitting the present moulded and carved sills and corbels.

On drawing No. 19, over the Ground Story window, the material is noted to be plaster, it should be marble.

The architraves of tablets over the foot of West Vestibule stairs will be moulded, as shown, but continuing on the under side of tablets. The carving of these architrave mouldings will be omitted; also, the sills and corbels.

There will be no plaster panel over elevator fronts, Ground Floor, as noted on drawing. The opening for wrought iron grill will be square form, and the architrave around opening to be similar as the architrave around door opening.

The architraves of doors, windows and elevator fronts on Office Floor will be moulded, as shown on drawings, but the window architraves to continue on the underside, omitting the present moulded and carved sills and corbels.

Omit entirely the architraves, sills and corbels forming panels or blind windows under West Vestibule stairs, the plain wall surfaces remaining to be radial jointed with the arch over. The clock window and window opposite in County Assessors' Office to have their architraves continue on the under side, same as specified for the other windows on Ground Floor.

The corners of the jambs and soffits forming arches under the West Vestibule stairs will not be beaded, as shown, but plain, and so formed with the faces as to disguise the connecting joints.

The architrave moulding of arch under East Vestibule stairs, Ground Floor, will be omitted entirely, forming a plain arch, the soffit and faces to be so formed as to disguise the connecting joints.

Except under all piers, pilasters, newels, balustrades, stair wall balustrades and small columns on stair newels, omit all moulded bases on floor plinths and substitute for same plain sub-plinths, as shown between tower pier moulded bases on drawing No. 17. To be the same kind of marble as moulded bases. There will be no marble sub-plinths or moulded bases in Judiciary, Public, Communicating or

Witness Corridors, from the inner corridor sides of elevator fronts and opposite, and from the inner corridor sides of wide pilasters on Law Library side of Rotunda, as all of these sub-plinths and moulded bases will be scagliola. All floor plinths of columns will be marble, the moulded bases and sub-plinths scagliola.

All moulded marble bases will be bed jointed between the top fillet and torus, the fillet to be cut on the course above, including the moulded bases of balustrades and newels. The top of sub-plinths to be level with this bed joint.

Wherever marble comes in connection with plaster caps the neck moulding or astragal must be marble, not plaster, and the lower fillet of balustrade rail and newel cap to be cut on the baluster course of balustrades and newels.

Provide a flat, vertical surface of marble to receive the U. S. Mail Chute, twelve inches wide, and extending from the top of the Mail Box, six feet above the finished Ground Floor, to a point four feet six inches above the finished floor of Court Room Story, this surface to be plumb from top to bottom and flush in both stories.

The stairs in North and South Lobbies and West Vestibule running from Ground Floor to Office Floor, and the stairs in East Vestibule running from the Ground Floor to Court Room Floor, to have marble treads and platforms two inches thick, each tread and platform to be in one piece, where possible, and thoroughly secured to the iron work.

The marble must be best quality Italian marble used for this purpose.

All plain floor bases or plinths throughout the Vestibules, Lobbies, Rotunda and Corridors on Ground, Office and Court Room Floors, all plain stair and floor bases or plinths of balustrades on stair and floor sides only of balustrades, the plain bases or plinths of all newels on stair and floor sides only, and the risors of all stairs, to be best quality blue veined Italian marble.

All marble above these plain plinths or bases to be best quality English Veined Italian marble, including the moulded bases on the plain bases or plinths, and the well and stair facias, soffits and for all other work, except carving.

All marble for tablets and carving to be white Italian marble free from veins and unclouded.

Wherever flush wall panels are shown, contrasting shades of specified marble will be used.

Bidders are requested to submit with their proposals for the above specified Italian marble work two alternate proposals, one for Vermont marble and the other for Tennessee marble. All to conform to the original drawings and specifications and these suplementary specifications so far as they are in conformity with the changes herein specified.

The proposals on Vermont and Tennessee marble, for stair treads and platforms must be for the best used for this purpose, and for all carving, to be free from veins, and unclouded.

The proposal on Vermont marble for all work specified above to be blue veined Italian marble, must be for the best quality medium clouded Vermont marble, and for all work specified above to be English Veined Italian marble must be for the best quality light clouded Vermont marble.

The proposal on Tennessee marble for all work specified above to be blue veined Italian marble, must be for the best quality dark Tennessee marble, and for all work specified above to be English Veined Italian marble, must be for best quality pink Tennessee marble.

All marble must be the best of their respective kinds, and bidders are requested to submit with their proposals samples of the marbles specified.

OTHER MARBLE WORK.

The contractor for the general marble work will furnish ready for setting, and the plumbing contractor will do all setting of all marble for partitions in Lavatory off Commissioners' Court Room and in the Jurors' Retiring Room off Court Rooms Nos. Two and Three, the water closet stall partitions, fronts, backs and ends, and the urinal partitions, backs, ends and bases. The plumbing contractor will furnish and set the marble backs, slabs and aprons necessary for all wash basins, including the wash basin in Power Station. The marble for this work to correspond with the marble specified for stalls, wainscoting, etc., otherwise to be as originally specified.

The plumbing contractor shall furnish and set all brass work for the above mentioned marble work. This brass work to be as called for in the original specifications.

The mirrors and mirror frames and the drinking fountains will be omitted entirely.

There will be no marble in the Power Station, except the above mentioned wash basin.

PLAIN PLASTERING.

The contractor for this portion of the work is referred to the specifications for Constructional Terra Cotta and Plastic Work, where changes are specified that will affect this portion of the work.

The original drawings and specifications are to remain the same so far as they conform to the changes specified herein.

All one by one inch tees for this work, and not otherwise specified, to be furnished and secured in place by the contractor for this work, as noted on drawings Nos. 53 and 55.

All metal lath for this work to be furnished and secured in place by the contractor for this work. He is also to furnish all necessary and required iron furring not called for in the specifications for structural iron work.

The spacing of iron beam brackets for lathing, where required, may be eight inches on centers, instead of twelve inches, as originally specified.

The attic Jury Room and closet ceilings are tile, not lath. All metal lath mentioned in iron work, and necessary for this work, must be furnished and secured in place by the contractor for this work.

Omit the one inch by one-quarter inch strap on all external angles, as called for in the original specifications.

Plaster the Private Staircase Hall. This was originally intended to be scagliola.

Omit the plastering in room in attic at attic stair landing and the walls and ceilings of stair way leading from this attic room to balcony over and around inner dome.

Omit the plastering on partitions on attic side of attic Jury Rooms and closets; also, the attic side of partitions of Law Library alcoves.

Omit the plastering on partitions in attic between Court Rooms.

Wherever cement architraves are omitted around doors, windows, etc., the corners are to be neatly rounded one inch in radius, and wherever cement work, ornamental plastering or scagliola is omitted it must be replaced with plaster by this contractor, unless otherwise specified, and all this plastering must conform to the requirements of the original specifications.

There will be no plastering in the Power Station.

PLASTIC WORK.

Excepting the false beams in the ceilings of Vestibules, Lobbies, Rotundas, Corridors, the Four Court Rooms, and the Commissioners' Court Room, ceiling of Grand Staircase Hall and the upper ceiling of the Private Staircase Hall, all false beams in the ceilings of Ground, Office and Court Room Storys, will be omitted. This includes the false beams forming panels in the offices of the Sheriff, County Superintendent, Township Trustee, County Assessors and the Assembly Room on Ground Floor, the Auditor, Treasurer, Clerk and Recorder on Office Floor, and the Law Library on Court Room Floor.

In all ceilings, excepting the ceilings of the above mentioned Vestibules, Lobbies, Rotundas, Corridors, the Four Court Rooms and the Commissioners' Court Room, ceiling of Grand Staircase Hall and the upper ceiling of the Private Staircase Hall, all cornices and mouldings, both plain and ornamented, beam soffit panels, beam brackets, and all ornamental panels, mouldings and enrichments of every kind will be omitted from the ceilings in all rooms, offices, lavatorys, passage ways, and closets, on Ground, Office, Court Room and Attic Floors will be omitted.

Excepting in the above mentioned Vestibules, Lobbies, Rotundas, Corridors, the Four Court Rooms and the Commissioners' Court Room, omit in all rooms, offices, lavatorys, passage ways and closets on Ground, Office, Court Room and Attic Floors, all cement architraves of door, window and alcove openings, entablatures, brackets, false piers and pilasters, and the bases and capitals of all columns, and omit entirely the false column in Treasurer's Office. The remaining columns in offices to be finished in cement, not scagliola, and omitting the bases and capitals. There will be no diminution or entasis of shafts; they to be finished perfectly plumb and of the required diameter. By omitting the false beams forming the entablatures over these columns, the columns will necessarily be greater in height than shown on drawings.

See specifications for Constructional Terra Cotta and Plain Plastering..

The ceilings and cornices of Vestibules, Lobbies, Rotundas, Corridors, the Four Court Rooms and the Commissioners' Court Room, ceiling of Grand Staircase Hall and the upper ceiling of the Private Staircase Hall will remain as they were originally intended.

The marble shown on drawings around and under, between aprons and base, of outside windows in Commissioners' Court Room and the marble jambs and soffits of doors and windows shown on drawings for the Four Court Rooms will be changed to scagliola. Other changes in the Four Court Rooms are specified below.

The columns in the above mentioned Vestibules, Lobbies and Rotundas will remain as originally intended. The bases, including sub-plinths on marble floor plinths, and the shafts, including astragals, will be scagliola. The capitals of all columns, piers and pilasters will remain as originally intended, the astragal of piers and pilasters being marble, and those of columns being scagliola, as above mentioned.

Wherever plaster caps come in connection with marble, the astragal will be marble, not plaster.

There will be no plaster panel over the Ground Floor elevator fronts, as noted on drawing. This panel will be made up of marble.

The caps above astragal of pilasters at Ground Floor starting points of East Vestibule stairs, the brackets over these caps and those opposite at intersections of beams between these brackets and the beams spanning between tower and arch under said stairs and the soffits of these beams to be plaster instead of marble, as originally intended and shown on drawings Nos. 18 and 23. The soffits of these beams to be paneled and ornamented similar to the corresponding beams above. The facias of all these beams, the brackets on said arch and the brackets under landings between Office Floor and Court Room Floor to be marble, as originally intended and shown on said drawings.

The spandrel panels formed by the intersection of ceiling plaster beams, stair soffit marble beams and ceiling plaster beams, and the circular balustrades in both of North and South Lobbies and East Vestibule stairs shown on ceiling plans of stairs on drawings Nos. 22 and 23, will be plaster, and not marble, as one of them is noted.

On Court Room Floor between marble sub-plinth course and plaster entablature on Court Room sides of Rotunda, between marble tower piers and on Law Library side of Rotunda, between marble pilasters opposite marble tower piers,

all work therein, consisting of the work around doors entering Court Rooms Nos. Two and Three and Law Library,
fluted pilasters at each side and tablets over the doors, window and tablet architraves, sills and corbels, and all plain
surfaces, is changed from marble to scagliola, excepting the
capitals above astragals of pilasters, the moulded inscription
panels in tablets over doors, the brackets, cornices and pediments over doors and the ornamental work in panels between
cornices and architraves over doors, which will be plaster.
Omit the enrichment of sills and corbels under tablets and
windows.

On drawing No. 17 the frieze of entablature is noted in one
place to be marble. It should be plaster.

The arch window over East Vestibule stairs will be raised
about two feet, as specified in stone work. The scagliola
base of pilasters will not continue. The window stool and
apron of the window will be marble, to stop or return on
themselves between the scagliola pilasters on each side of the
window. The additional surface between the marble plinth
course under pilasters, and the apron under window stool
will be scagliola, formed in panels. The raising of this arch
window will crowd the eagle nearer the balcony cornice
above and into the ornamentation under this cornice.

The two members of the small ornamented moulding between plaster panels and architraves over entrances to Court
Room Floor Corridors will be scagliola, instead of marble,
as indicated by witness lines on drawings Nos. 17 and 18.
Omit the enrichment of this moulding.

In from the sides of elevator fronts, and opposite, and in
from the sides of wide pilasters, and opposite, on Law
Library side, in Judiciary, Public, Communicating and
Witness Corridors, between the marble floor plinth and
plaster cornice, all wall surfaces, pilasters, mouldings and
architraves will be scagliola, excepting the plaster panels
over doors. Omit the enrichment of mouldings on pilaster
caps and on mouldings over doors.

There are to be architraves to Communicating Corridor
doors.

The caps of wide pilasters forming jambs of entrances to
Judiciary Corridors will be marble, instead of plaster, as
noted on drawing. Omit enrichment of mouldings. The
soffits over these caps will be marble, except the panels,
which will be plaster

The architraves between balcony and inner dome will also be scagliola, instead of cement, as noted on drawing.

Omit the scagliola in Private Staircase Hall, full run.

Do all trimming around Mail Chute in a neat and careful manner.

Court Room No. One is to remain as originally intended, with the following exceptions: The fluting of columns to be omitted, and the plinths under bases of columns to be scagliola instead of cement, as noted on drawing. The columns and pilasters are to be scagliola from the top of marble floor plinths to and including the fillet under plaster capitals, and to bead mould of pilaster capitals. The architraves of doors and windows to be scagliola instead of cement, as noted on drawing, the enrichment of mouldings of these architraves to be omitted. Omit the rosettes at the sides of door back of the Judge's Bench, the remaining portion of this work at the sides of the architrave to be scagliola. The door in alcove opposite the Judge's Bench to be similar to entrance door shown, and the windows in alcove similar to the alcove window shown on drawing for Court Room No. Four, with plaster panels over all of them. The outer wall sides of Court Room to be the same as Communicating Corridor side.

Court Room No. Two is to remain as originally intended, with the following exceptions: The fluting of columns to be omitted, and the plinths under bases of columns to be scagliola instead of cement, as noted on drawing. The columns and pilasters are to be scagliola from the top of marble floor plinths to and including the fillet under plaster capitals and to bead of pilaster capitals. The architraves of doors to be scagliola instead of cement, as noted on drawings, the enrichment of mouldings of these architraves to be omitted. Omit the fluting of arch imposts and between the columns back of the Judge's Bench; also, the enrichment of mouldings of impost jamb panels, and the enrichment of mouldings of panel between imposts. Also, omit the modilions and panel between modilions under clock, leaving the impost cap course unbroken between the caps. All of this work from the top of marble floor plinth and base to and including the bead mould or astragal of impost cap to be scagliola, omitting the astragal beads. The work between this astragal and soffit of cornice and between the columns to be plaster. The doors, panels, etc., on Public and Communicating Corridor sides of this Court Room to be respectively the same as shown on drawing for Tower and Judiciary Corridor sides.

Court Room No. Three is to remain as originally intended, with the following exceptions: Omit the columns mentioned in note No. 2 on drawing and their corresponding pilasters and ceiling beams in alcove, and the festoons in frieze of cornice over these omitted columns. The panels of alcove ceiling to be widened to make up for the omitted ceiling beams over the omitted columns and pilasters. The remaining columns to be fluted as originally intended and shown on drawing. The plinths under bases of columns to be scagliola instead of cement, as noted on drawing. The columns and pilasters are to be scagliola from the top of marble floor plinths to and including the fillet under the plaster capitals and to bead mould of pilaster capitals. The architraves of doors to be scagliola instead of cement, as noted on drawing, the enrichment of mouldings of these architraves to be omitted. Omit the rosettes at the sides of door in alcove, the remaining portion of this work at the sides of the architrave to be scagliola, the console and balance of work over door to be plaster, as noted on drawing. Omit the enrichment of mouldings of arch impost panels and between the columns and pilasters back of the Judge's Bench; also, the enrichment of panel and enrichment of mouldings of panel between imposts. All of this work from the top of marble plinths to and including the fillet over and above the astragal of impost cap course to be scagliola, omitting the astragal beads. The work between the above mentioned fillet and soffit of cornice and between the columns to be plaster. The doors, panels, etc., on Public and Communicating Corridor sides of this Court Room to be respectively the same as shown on drawings for Tower and Judiciary Corridor sides, except that the projection of pilasters on the Public Corridor side will not be as shown on plan, but similar to that shown on plan for Court Room No. 2, thereby obviating the necessity of breaking the entablature and forming a pediment. Therefore the pedimental feature on Judiciary side of Court Room No 3 will not be repeated on the Public side, as noted on drawing for this Court Room.

Court Room No. Four is to remain as originally intended, with the following exceptions: The fluting of columns to be omitted, and the plinth under bases of columns to be scagliola instead of cement, as noted on drawing. The columns and pilasters too are to be scagliola from the top of marble floor plinths to and including the fillet under plaster capital, and to upper bead mould of pilaster capitals. The architraves of doors and windows to be scagliola instead of cement, as noted on drawing, the enrichment of mouldings of these architrives to be omitted. Omit the rosettes on door architrave back of the Judge's Bench and the enrichment under consoles. The consoles and balance of the work over this door to be plaster, as noted on drawing. The clock and sculptural work attached thereto will not be repeated on opposite end. The door in alcove opposite the Judge's Bench to be similar to the doors on each side of the Judge's Bench and the windows similar to the window shown on drawing, with plaster panels over all of them. The outer Communicating Corridor side of Court Room to be the same as the outer wall side.

Separate models must be made for the sculpture in each and every pediment, panel, spandrel and all other sculpture, as no duplicates will be accepted, and models of all important capitals, ornamented, etc., will be required.

All moulding to be done on the Court House premises, and placed in proper heighth and distance position for inspection.

The contractor for this portion of the work is referred to the specifications for Constructional Terra Cotta, Plain Plastering, and Marble Work where changes are specified bearing on this portion of the work. The original drawings and specifications are to remain the same so far as they conform to the changes specified herein.

All one by one inch tees for this work, and not otherwise specified, to be furnished and secured in place by the contractor for this work, as noted on drawings Nos. 53 and 55.

All metal lath for this work to be furnished and secured in place by the contractor for this work. He is also to furnish all necessary and required iron furring not called for in the specifications for structural iron work.

The spacing of iron beam brackets for lathing, where required, may be eighteen inches on centers instead of twelve inches, as originally specified.

All metal lath mentioned in iron work, and necessary for this work must be furnished and secured in place by the contractor for this work.

Omit the one inch by one-quarter inch strap on all external angles, as called for in the original specifications.

The scagliola, in such imitations as may be selected by the Architect, is to be manufactured on the Court House premises, and as follows: The bases in every case to be Keene's Superfine cement, with such proportions of coloring matter as will not injure its natural quality; the coloring matter to be strictly pure staining materials of their various kinds. This work to be colored at the building, made in sections except columns, to a thickness of at least one-half inch, the entire coloring and veining to run completely through. The sections are to be set in place, all joints to be carefully faced up, and when sufficiently hard, are to be smoothed down perfectly true and level, then all surfaces are to be hand polished with the various stones and water necessary for this purpose, as many times as may be required and requested by the Architect, and until the surface presents a brilliant polish, equal to a first-class piece of marble.

Absolutely no shellac, wax or chemicals will be allowed in the finishing of this work.

The necessary ground work for the application of the scagliola to be provided as the Architect may direct.

The contractor shall be required to absolutely guarantee the scagliola as to its standing and wearing qualities in every instance.

Bidders are requested to mention in their proposals the names of the sub-contractors they purpose doing this work, but the acceptance of any proposal by the Commissioners must not be construed by the contractor as an approval of any sub-contractor, as none but those thoroughly competent and reputable will be approved by either the Commissioners or Architect.

There will be no cement ornamental plaster work or scagliola in the Power Station.

MAIL CHUTE.

Furnish and erect, where located on plans, on a marble casing otherwise provided for, one U. S. Mail Chute and special receiving Box, the box to be located in the Ground Story and the chute connected to the box extending upward in a vertical line through openings in Office and Court Room Floors, otherwise provided for, to a point four feet six inches above the finished floor in Court Room Story, with openings for mail in Ground, Office and Court Room Storys.

The Mail Chute to be U. S. standard pattern, and the Mail Box to be cast bronze, in accordance with the design shown on drawing sheet No. 71. To be finished in the best manner and supplied with such devices for preventing injury to the mail, etc., as are required, the whole apparatus to conform in every way to the rules and regulations of the Post-office Department, governing the construction and location and arrangement of U. S. Mail Chutes.

All exposed metal work connected with the Chute to be finished in copper deposit on iron, and the face to be best plate glass furnished by this contractor.

The Mail Chute complete to be equal in every respect to those manufactured by Cutler's Manufacturing Co., Rochester, N. Y.

The contract to be approved by the Postmaster of Fort Wayne, Ind., and the entire work to be done in a first-class workmanlike manner, to the entire satisfaction of the Commissioners and Architect.

The price stipulated in the original specifications for this work will be no longer considered.

WOOD WORK.

Omit all grille work in the frames of large transom arches in Court Room Floor Rotunda and Law Library; also, in the frames of windows between tower piers, Court Room sides.

Omit metal bars in windows at side of doors. This grille work in the windows to be made up of white oak, to be constructed for securing glass same as other sash.

Omit the panels and the ornamentation in panels of transoms over all doors, both sides. Also, omit the egg and dart and bead ornamentation of the mouldings of these transoms on all room sides. Omit carving in transoms over doors in East Vestibule, Ground Floor, to be left in plain moulded panel.

Omit carving in panels of pilasters at side of doors on all room sides. Also omit the enrichment of the mouldings of these panels, the enrichment of the mouldings of the panels under these pilasters and under the windows at side of doors and the enrichment of the mouldings of the door panels on all room sides, excepting Court Rooms.

Omit iron grille in panel of doors in Court Room No. One.

Instead of a door in County Assessor's Room opposite Clock Room window there will be a window similar to said Clock Room window.

The sash in Court House and the sash and doors in Power Station will not be veneered, as originally specified, but to be solid, and of selected clear white pine.

There will be no glazed or screen partition between the machinery and boiler rooms of Power Station, nor will there be any inside finish.

The cornice of Power Station will be galvanized iron instead of terra cotta, as originally specified.

The contractor for this portion of the work connected with the Court House shall furnish and secure in position approved awning fasteners for all east, south and west windows of the Court House. This was originally included with the structural iron work.

The closet and coal bin partitions in the Power Station will be constructed of heavy plank and posts, the coal bin partition to be arranged and constructed for raising and lowering in sections.

HARDWARE.

The contractor for the ornamental iron work shall furnish and place all hardware for the front doors of the Court House, as specified in Ornamental Iron Work.

GLASS AND GLAZING.

The mirrors for Lavatorys have been omitted. There will be no glazed partition between machinery and boiler rooms of Power Station.

PAINTING.

In changing the inside of sash from oak to pine, it will be required to stain and grain the pine to imitate quartered white oak. To be done in the very best and first-class manner, and finished as originally specified.   •

The galvanized iron cornice, roof ridges, hips, etc., to be painted and sanded two coats in addition to priming.

There will be no inside finishing work in the Power Station. All white pine therein to be painted as originally specified.

All pipe railing, iron ladders, area covers, iron thresholds, etc., in Court House and Power Station and Stack to be painted as originally specified for other iron work.

All finishing tints to be approved.

ROUTE OF TUNNEL.

The drawing shows that portion of the Tunnel as being on the east side of the sewer, thereby conflicting with the specifications as described on page 181. The route of the Tunnel should be as described in the specifications, the Tunnel crossing the main sewer on Calhoun Street in a line with that portion of the Tunnel "C-J," as shown on the drawing, to a point bringing the center line of the Tunnel, as it runs north and south, ten feet two inches west of the cener line of the main sewer, which also runs north and south on Calhoun Street. In addition, the Tunnel should be run from the south side of Power Station further north than figured on drawing No. 1. Thence turning directly west and extending into the Power Station up to the wall dividing the boiler from the engine room. The center line of the Tunnel entering the Power Station must be approximately thirty feet from the inside line of south wall of the Power Station. See specifications for Cable Racks and Steam Pipe Hangers And Supports.

CABLE RACKS.

The iron cable racks, as called for to be provided by the Tunnel Contractor, are under the revised specifications, to be provided by the Contractor of Electrical Construction, but the Tunnel Contractor shall place such racks and supports as furnished by the Electrical Construction Contractor, the arrangement and location being such as laid out by the Contractor undertaing the electrical portion of the work.

PIPE HANGERS AND SUPPORTERS.

The Tunnel Contractor will not furnish any of the iron hangers, as called for in the original specifications, neither will he be required to furnish the roller bearings, plates or saddles. The above are to be furnished by the Steam Heating Contractor. The Tunnel Contractor shall, however,

provide all brick supporting piers and properly place all
hangers and roller bearing supports, complete, including such sliding plates as the drawings call for, the above to
be placed at such points as may be directed by the Piping
Contractor.

ELECTRICAL CONSTRUCTION.

In addition to the labor and material, as provided for under the original specifications for Electrical Construction,
this contractor will be required to furnish all cable racks, as
called for on page 183 of the original Tunnel specifications,
provisions having been made above referring to that portion
of the work for such; racks or supports to be properly placed
in position by the Tunnel Contractor, the location of same to
be indicated by the Electrical Cnstruction Contractor.

STEAM HEATING.

Under the heading of Steam Heating, there will be included
in addition to the material and labor called for in the original specifications, the furnishing of all hangers, roller bearings, saddles and sliding plates, all as called for on page 183
of the original Tunnel specifications. The Tunnel Contractor will properly attach and set all supporting racks and
plates as provided above in the revised Tunnel specifications.

REFRIGERATING APPARATUS.

The entire refrigerating apparatus and drinking water system, including filter, ammonia compressor, cooling tank,
circulating pump, electric moter drinking fountains, piping
and valves for drinking water system, as called for in the
original specifications, is, under these revised specifications,
to be omitted from this contract.

PUBLIC PASSENGER ELEVATOR.

The south public passenger elevator, including all equipment in connection with the same, as called for in the original specifications, shall be omitted from this contract, with
the exception that the guide posts, sheaves and beams and
foundations shall be put in place, as called for in the original
specifications, ready to receive the remaining equipment of
elevator whenever same is installed. The elevator front and

the elevator door opening mechanism shall also be put in place.

PIPING OF POWER STATION.

The specifications covering the piping of Power Plant, beginning on page 158 of the original specifications and concluding on page 164, will remain the same, with the exception of omission of that portion of the specifications beginning on the top of page 159—"Contractor shall connect each of the main five inch angle valves, etc."—and concluding with paragraph at the top of page 160, and also the omission of that portion of the paragraph at the bottom of page 161, which calls for steam connection to pumps to be taken directly from main steam header, for which portions, the following will be substituted:

Contractor shall connect each of the main six-inch angle valves, as elsewhere specified, to the boiler nozzles and connect same by six-inch pipe to a ten-inch main steam header; pipe to rise from the boilers, run over and into the top header, which will be located above and toward the front of the battery of boilers, connecting with a second ten-inch steam header, substantially as shown on the drawing.

A six-inch gate valve shall be placed in each steam pipe, connecting with main header close to the header, in addition to the angle valves placed close to boiler. In the ten-inch header, running north and south, are to be placed two ten-inch gate valves. The main header at boilers to be fitted with three ten-inch by ten-inch by six-inch double sweep tees. The west end of header to be closed by blind flanges. The six-inch pipes from boilers to pitch toward the header.

The steam pipes on each of the engines to be five-inch; pipe lead out of the top of the ten-inch leader, running north and south, by long bend fittings and connected to a main five-inch "Chapman" valve, and thence to special five-inch copper bends of thirty-inch radius, connecting to the vertical separators, as elsewhere specified, separators to be fastened directly to throttle of engines.

The contractor will furnish one four-inch "Clevauc" pressure reducing valve capable of delivering any pressure between ten pounds and a vacuum equivalent to two pounds.

There shall also be a header of four-inch pipe running

across the top of boilers and connected into each of the three steam nozzles by two-inch nipple and valve. Near the east end of this four-inch header shall be placed a one-inch tee for steam connection to injector, and just beyond this tee the pipe shall divide and run by one and one-quarter-inch pipe and valves to the throttle valves of the various pumps. A cross connection shall also be made from this four-inch header, without reducing, into the live steam connection at reducing valve; said connection to be provided with valve at each end.

The connections for the steam pipe to the pumps to be made with double elbow to each pump. The main exhaust pipe from each engine to be six inches in diameter, located in position substantially as shown on the drawing. This exhaust piping to be "light," wrought iron, lap welded pipe. Piping to run from each engine to a twelve-inch exhaust header, running to the feed water heater. Contractor to place an oil separator as indicated in each exhaust pipe. The separators to be provided as elsewhere specified.

The exact location of pumps, heater, catch-basin and trap will be determined after contracts are let and apparatus selected, but is intended to be essentially as shown on the drawing.

The two main ten-inch steam headers are to be drained at at least three points, all connected together into a trap of proper size, which shall in turn deliver into the feed water heater. Valves to be placed at each connection to header.

Before the contractor begins work, he shall submit to the Architect for approval, complete piping plans, showing the sizes of all pipes and fittings.

SEPARATORS.

The three vertical separators called for on page 157 of the original specifications are to be five-inch instead of six. The three horizontal separators are to be six-inch instead of seven. The specifications covering separators, otherwise to remain the same.

BOILER BRICK WORK.

In place of the outside walls of the boiler setting being of white enameled brick, as called for on page 152 of the specifications, the outside walls shall be faced with selected red face brick. All other brick work in connection with the boilers to remain as called for in the original specifications.

FIXTURES IN POWER STATION.

The fixtures in Power Station shall be as called for in Plumbing specifications of Power Station and as shown on revised plans of Power Station.

TELEPHONE SYSTEM.

The Telephone System called for in the original specifications will be entirely omitted from this contract.

CONSULTING ENGINEERS.

Messrs. Pierce & Richardson, Chicago, Ill., are the Mechanical, Electrical, Heating, Ventilating and Sanitary Engnieers on this work for the Architect, B. S. Tolan, Fort Wayne, Ind.